Textile Science and Clothing Technology

Series Editor

Subramanian Senthilkannan Muthu, SgT Group & API, Hong Kong, Kowloon,
Hong Kong

This series aims to broadly cover all the aspects related to textiles science and technology and clothing science and technology. Below are the areas fall under the aims and scope of this series, but not limited to: Production and properties of various natural and synthetic fibres; Production and properties of different yarns, fabrics and apparels; Manufacturing aspects of textiles and clothing; Modelling and Simulation aspects related to textiles and clothing; Production and properties of Nonwovens; Evaluation/testing of various properties of textiles and clothing products; Supply chain management of textiles and clothing; Aspects related to Clothing Science such as comfort; Functional aspects and evaluation of textiles; Textile biomaterials and bioengineering; Nano, micro, smart, sport and intelligent textiles; Various aspects of industrial and technical applications of textiles and clothing; Apparel manufacturing and engineering; New developments and applications pertaining to textiles and clothing materials and their manufacturing methods; Textile design aspects; Sustainable fashion and textiles; Green Textiles and Eco-Fashion; Sustainability aspects of textiles and clothing; Environmental assessments of textiles and clothing supply chain; Green Composites; Sustainable Luxury and Sustainable Consumption; Waste Management in Textiles; Sustainability Standards and Green labels; Social and Economic Sustainability of Textiles and Clothing.

More information about this series at http://www.springer.com/series/13111

K. Murugesh Babu · M. Selvadass ·
Megha Shisodiya · Abera Kechi Kabish

Abstract Pattern Illustrations for Textile Printing

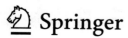 Springer

K. Murugesh Babu
Department of Textile Technology
and Research Centre
Bapuji Institute of Engineering
and Technology
Davangere, Karnataka, India

Megha Shisodiya
Fashion Designer
Label—MeThinks
Delhi, India

M. Selvadass
Department of Fashion and Apparel Design
Garden City University
Bengaluru, Karnataka, India

Abera Kechi Kabish
Scientific Director
Bahir Dar University
Bahir Dar, Ethiopia

ISSN 2197-9863   ISSN 2197-9871   (electronic)
Textile Science and Clothing Technology
ISBN 978-981-16-5977-5   ISBN 978-981-16-5975-1   (eBook)
https://doi.org/10.1007/978-981-16-5975-1

This Springer imprint is published by the registered company Springer Nature Singapore Pte Ltd.
The registered company address is: 152 Beach Road, #21-01/04 Gateway East, Singapore 189721,
Singapore

Foreword

This book is for textile designers and for everyone curious about the integration of graphic design with textile surface printing. The book is also a blessing for fashion designers who want to fully integrate graphic design with textile fabrics. Designers working on how abstract graphic designs with intense color palette range will work on different types of fabrics will also benefit enormously from this book.

The book embraces you with knowledge about different types of abstract designs. It provides lovely illustrations of abstract designs that can be used directly for textile printing and also can act as inspiration or motivation for the development of new designs. Abstract designs represent an accurate depiction of visual reality and use shapes, colors and forms to achieve its effect. It is also applied to art that uses forms, such as geometric shapes, which have no source at all in an external visual reality.

Color makes a design come alive and is an important parameter that is noticed by the customers. It can attract attention, set a mood and even influence our emotions and perceptions. Humans are trichromats and have three distinct types of receptor cells in our retina, each being sensitive to different light properties mainly to colors like red, green and blue. Hence, humans can see different colors because of these three little receptors. This book provides illustrations that show the importance of color and color combinations with bright, warm and dull colors. Different colors can create a different mood for a design. Mood means the feeling we get when we look into the design. We can create the mood by selecting warm or cool colors that remind us of the emotions that we want in our design. The book shows flawless illustrations with great harmony between the diverse shapes and overall color combinations.

All the illustrations in this book are explained briefly. The illustrations can also be used in other areas like wallpaper design, packaging design, ceramic design and many more. I hope the work will inspire you and that you gain a great deal of pleasure from this book.

Davangere, India   K. Murugesh Babu
Bengaluru, India   M. Selvadass
Delhi, India   Megha Shisodiya
Bahir Dar, Ethiopia   Abera Kechi Kabish

Introduction

Abstraction is a way of viewing and perceiving the world around us. Abstraction often originates from reality with a shift in emphasis from complete entity to the finer details. Abstraction, like a representative art, is an expression of the state of mind and passions of the artist. The features of the landscape are symbolized using the basic design elements with the emotions and perceptions of the artist articulated by the choice of colors.

Abstraction can be embedded with deeper spiritual meanings and can inspire the viewer to see beyond oneself. There is no precise formula or recipe for the creation of abstract art. There are some rudimentary guiding principles to commence the journey of abstraction with more than one route to reach the destination. Abstraction requires the application of all the fundamental and foundational elements of art along with much planning, thought and skill.

Abstract art can be non-objective or figurative. Non-objective abstraction involves the use of the elements of art like lines, shapes, colors, etc. Figurative abstraction uses abstracted backgrounds, animals, plants, flowers, etc. in the form of essential shapes with a modified outlook.

The principles of design are the guidelines an artist must follow to create a sophisticated and exceptional artwork. An artwork is evaluated by the arrangement and organization of the design elements. A few prominent principles to be applied in the design briefed here.

1. Emphasis: Emphasis is the pivotal point in the art that grabs the attention of the viewer. The artist may emphasize his ideas through variation in contrast or by the use of complementary colors or by adding texture.
2. Balance: The visual stability attained in the design by the placement of elements with equal weights about the central axis. The weight can be in the form of color, size or texture.
3. Contrast: Use of conflicting or dissimilar elements to emphasize and highlight the designer's theme and enhance the interest of the viewer.

4.   Repetition: Repetition adds pace and energy to the artwork. Repetition need not always mean sameness. The dimensions, shades and other aspects can be varied to augment interest.
5.   Proportion: The dimension and weight of design elements in an artwork with respect to other elements in the design.

The design elements in an art are the basic building blocks of art. Hence, a comprehendible knowledge about them is essential. They add meaning to our work and help assess them. A brief description of the basic design elements that are an indispensable part of an art is provided here.

1.   Line: A line is the most basic element of art. Type of line can vary in terms of its depth, weight, length, sharpness, etc. Lines may be straight or curving and frequently articulate the passions, state of mind and the personality of the artist.
2.   Shape: Shapes in an artwork may be organic, geometric or derived. Organic shapes help create more realistic arts while geometric shapes take one of the predefined forms. Arts and abstractions may also be developed using shapes derived from a plant, flower, etc. The variety, size and placement of shapes within a composition add curiosity to the design.
3.   Value: The term value in art and design refers to the value of light. By varying it, the artist can add contrast to the artwork thus making it more intriguing. A viewer's attention is drawn toward areas with higher contrast and that helps the artist to project what he thinks is important.
4.   Space: It is not only the object that brings out the ideas in a design but also the space around it. Not every area of the design has to be filled. Spaces between the designs add energy, emphasis, contrast and variety to the design.
5.   Color: An elementary understanding of color theory goes a long way in helping the artist with the choice of colors. The colors in an art provoke an emotional response from the viewer. Each color is associated with an emotion or state of mind. For instance, red is associated with aggression and danger, yellow with alertness and confidence, blue with balance and gloominess, etc. Stunning and exceptional patterns can be developed by varying the hue, value and intensity of the color.
6.   Texture: Texture creates interest in an art work. Texture may be added to the design by a variety of methods from the choice of material to the direction of brush stroke.

Abstraction has been predominant in all forms of arts and designs including textile designing. Contemporary abstraction in textile designs saw its beginning and growth with the era of modernism. With the onset of digital printing, printing abstract patterns on textiles have become a viable alternative.

Textile printing is a process where designs can be printed onto the fabric. The conventional method of textile printing involves the use of printing rollers and flat screens whose number is limited by the number of colors present in the design. Digital textile printing offers greater scope for printing complex designs directly onto the treated fabric without restriction on the number of colors. It offers greater design flexibility for printing complex designs and color combinations in production and has increased productivity. Digital textile machines can be used to print small batches with quality print and economy. Digital textile printing is a sustainable solution for meeting the increasing demand with minimal use of resources. It comes with the indispensable benefit of inexhaustible shades of colors and accurate translation of design details. Multicolor printing by conventional printing technique needs the use of separate templates for each color thereby increasing the cost and time consumption. Digital printing technique offers complete control over color enabling exact replication of colors from the original object including the intricate and delicate graphic details. The assortment and splendor of shades and colors we get from digital printing are supreme. The precision in digital printing is mainly attributed to the transfer technique employed, which also minimizes the intake of dyes and the menace of color setting off. Digital printing has eliminated the constraints on the maximum limit on colors and printing length. The problems due to overlaps in printing by conventional techniques can be overcome in digital printing with no restrictions on repeat size.

Abstract Pattern Illustrations for Textile Printing is a pattern book with the collection of 250 abstract illustrations. The book shows flawless abstract illustrations with great harmony between the diverse shapes and overall color combinations. The designs that can be used for textile printing also can act as inspiration for the development of new designs. The illustrations are all worked on different geometrical shapes forming one single illustration or a print. The basis of each pattern is a motive or a shape that is translated, reflected, repeated or rotated to create a new digital version of the shape that is then aligned with or superimposed upon itself. Most of the motives are geometric, inclusive of spheres, circles, squares, hexagons, etc. or they are representative of a shape such as flowers, leaves, cones, alphabets, arrows, horns, figures, loops, spirals, chevrons, vines or mesh. A number of motives we utilized here are inspired by nature's floral diversity and include modern abstract digitalized shapes. The patterns can be used for printing on textile fabrics using various commercially available printing systems such as digital and transfer textile printing machines. These designs will enable us to explore and develop new design inspirations with a clear sense of direction.

Design decisions are made at every stage from the manufacturing process to production, selecting a color to the kind of print and finishes to complete a product. During the design process, a designer always looks for something new and fresh to create a unique product. Being aware of this demand for authentic work in the design industry, we wanted to create a primary design and print that could directly be used by designers to develop fashionable textiles, home furnishing, product and lifestyle design.

This book is intended for a broad spectrum of readers worldwide, ranging from undergraduate/post-graduate students, designers and technical staff working in the field of textile and clothing, to designers and product development staff working in designing and printing in general. The designs are organized into 10 chapters with all the illustrations explained briefly. The illustrations can also be used in other areas like wallpaper design, packaging design, ceramic design and many more.

Contents

About the Authors

Dr. K. Murugesh Babu has more than 30 years of teaching and research experience in the field of textile science and technology. Dr. Babu, a doctorate from IIT, Delhi has rich experience in teaching and research in the areas of silk technology, textile dyeing, printing and finishing, natural fiber processing, textile fiber composites, textile polymer science and was awarded the prestigious Commonwealth Fellowship by the Association of Commonwealth Universities, the UK in 2014. Under this fellowship, he was invited as a Visiting Professor by the University of Maroua, Cameroon, Central Africa to establish and develop the Department of Textiles. Dr. K. Murugesh Babu has published more than 100 research papers in various national and international textile and polymer-related research journals. He has also contributed 10 Book Chapters in textbooks published from Wood Head Publishing and Elsevier. He has published two textbooks on Silk.

Dr. M. Selvadass is working as an Assistant Professor at Garden City University and obtained his doctoral degree in the field of Textile Technology from Visvesvaraya Technological University, Belagavi, Karnataka, India. He has 16 years of industrial experience working in reputed garment manufacturing companies and buying houses. His field of specialization is apparel and printing technology. In the printing sector, he has worked extensively in the area of flexographic, rotogravure, screen and digital textile printing. He has worked in different departments like process development, involved in the development of designs and preproduction activities like development of flexographic design plate, gravure cylinders and screens for printing. His experience in the production department involves printing of design onto the material and finishing of the printed materials.

Megha Shisodiya has worked as a Design Academician for 5 years at Amity University, Khazani Women Polytechnic and Bangalore University, India and has 3 years of experience in Design Retail with Kimaya and Raghavendra Rathore. She has independently worked for NGOs and training women for work and career counselor for design aspirants. She is a fashion designer at Label—MeThinks.

Dr. Abera Kechi Kabish obtained his Advanced Diploma, Bachelor's Degree and Doctoral Degree from Bahir Dar University (Ethiopia) with a Master's Degree from Anna University (India). His main teaching and research experience is in the areas of textile dyeing and printing. He has published more than 12 research papers in various Scopus indexed international journals and has administered a number of research projects related to natural dyes and printing.

He has vast experience in leading higher education institutions in the capacity as Dean, Associate Vice President and Scientific Director. At present, he is working as the Scientific Director at the Ethiopian Institute of Textile and Fashion Technology, Bahir Dar University, Bahir Dar, Ethiopia.

Chapter 1
Abstract

Abstract art, whose origin is presumed to be the ancient cave paintings, saw a major breakthrough a century ago, inspired by the evolution of new ideas, philosophy and technology. The idea of art shifted focus from accurate figurative representations to the use of shapes, colors and gestures. Abstract means to separate one thing from another and involves geometric blocks of uniform shapes and colors. Abstraction is manifested around the world in art and can be seen as an imperative component of modern art. Abstract art is usually non-representative, and its interpretation is left to the imagination and experience of the individual. Abstract art usually has a moral or spiritual aspect attached to it that exhibits virtues like order, simplicity and purity.

Fabrics with abstract art have become an integral part of life, and every culture has some kind of traditional abstract textile design. Abstract art creates appealing and serene designs in textiles. In this chapter, designs with abstract art that can be printed on the fabric surface are presented. Designs with various shapes, colors, styles, themes and tones are presented. Inspirations from the natural world have been tailored to create aesthetic, colorful and artistic abstract designs (Figs. 1.1, 1.2, 1.3, 1.4, 1.5, 1.6, 1.7, 1.8, 1.9, 1.10, 1.11, 1.12, 1.13, 1.14, 1.15, 1.16, 1.17, 1.18, 1.19, 1.20, 1.21, 1.22, 1.23, 1.24, 1.25, 1.26, 1.27, 1.28, 1.29, 1.30).

K. Murugesh Babu et al., *Abstract Pattern Illustrations for Textile Printing*, Textile Science and Clothing Technology, https://doi.org/10.1007/978-981-16-5975-1_1

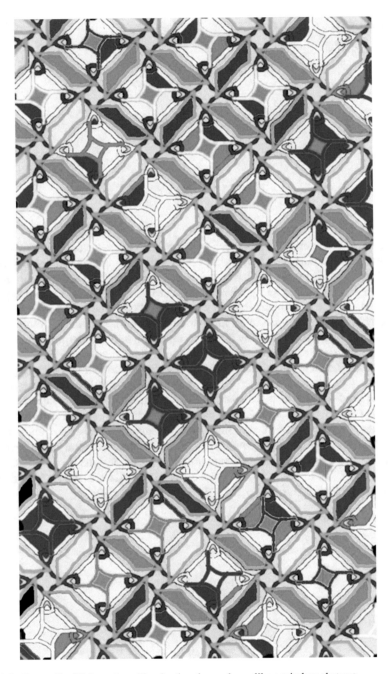

Fig. 1.1 Fenestella: Niche pattern illuminating the credence like a window abstract

Fig. 1.2 Bewitching hour: Simple yet enchanting floral pattern with peace and delightful abstract

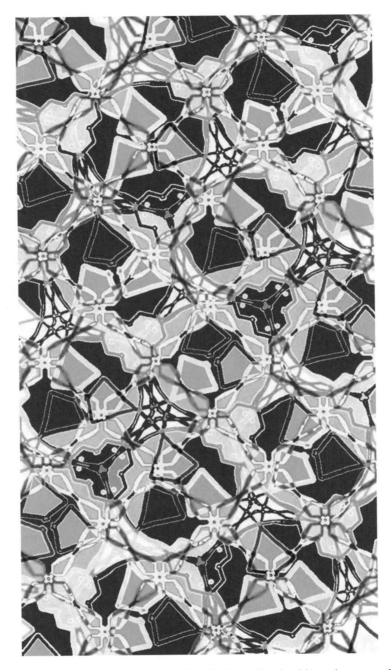

Fig. 1.3 Naval: Subdued blue tone pattern with a flashing outline in white posing as a starlit

Fig. 1.4 Formation: A lively expression created with structural base and contrast color scheme

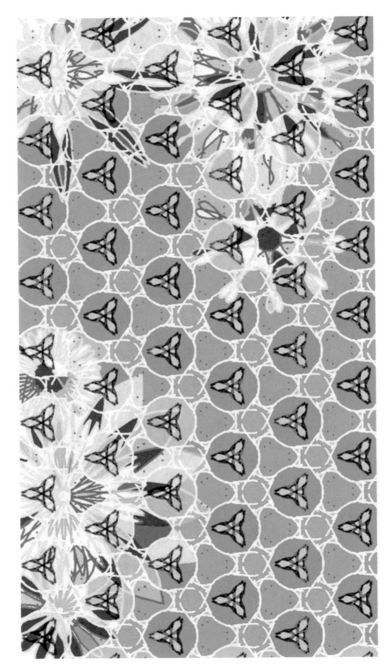

Fig. 1.5 Livable: Florals creating an illusion of a cozy and fresh pattern like a summer breeze

Fig. 1.6 Brilliant: Skillfully used shaped mortifying into well-acted floral mesh

Fig. 1.7 Gloom: Partials of blue prism peering into electrified motifs

Fig. 1.8 Upsurge: Twisted shapes showing a strong pattern with a self-design illusion

Fig. 1.9 Oriel: Intricate pattern curated delicately using spherical shapes in minimalistic way

Fig. 1.10 Temperate: Color codes denoting a mild combination creating a warm and balanced character as a texture

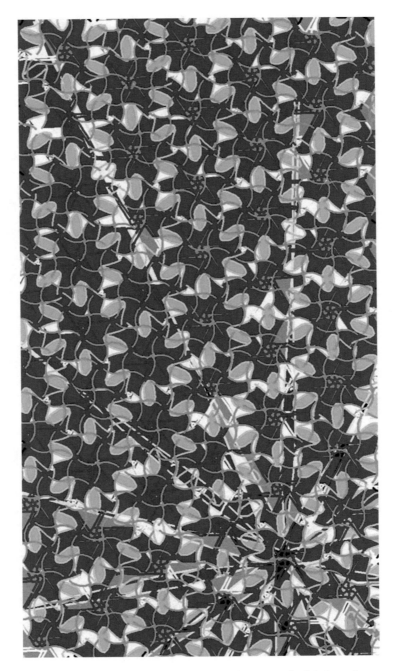

Fig. 1.11 Scribble: Carelessly distributed shapes drawn into a meaningful piece of texture

Fig. 1.12 Ellipsoidal: Transformation of spherical shapes through directional scaling into the design surface

Fig. 1.13 Canape: Small bent shapes held in a cracker format like a decorative surface

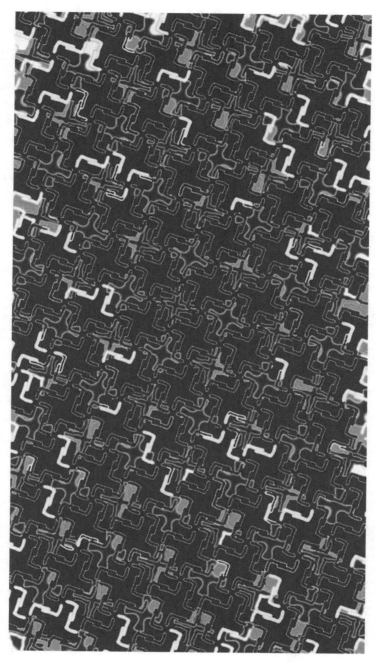

Fig. 1.14   Ultramarine: Grinded shades of blue spread across like pigmented geometrical shapes

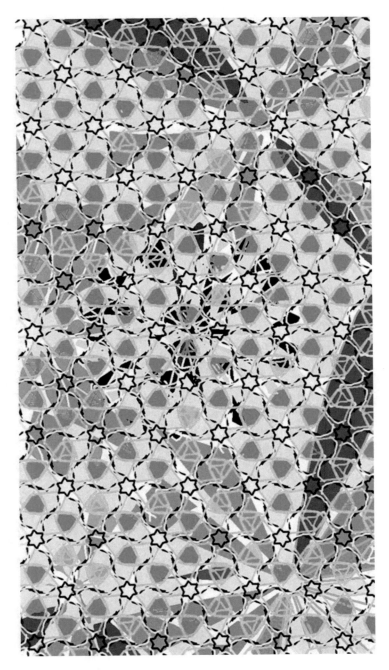

Fig. 1.15 Dash one: Diamond and star shapes in hush tone independently creating a remark of repeated abstract

Fig. 1.16 Framework: The supporting and delicate enduring shapes framing a structured pattern

Fig. 1.17   Blitheness: A carefree pattern placed in a happy spirit with repeated pointers

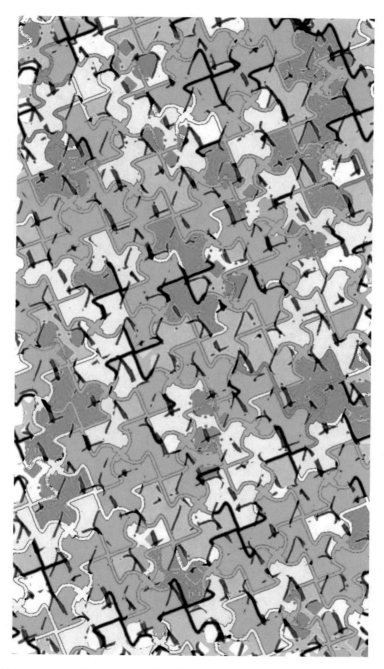

Fig. 1.18 Meticulous: Attention to motif detail creating a glazed and précised Lead

Fig. 1.19 Efflorescent: Migrated clover loops forming a blend of two floral shapes on the surface

Fig. 1.20 Eventide: An evening shaded geometrical structure in a diamond repetition pattern

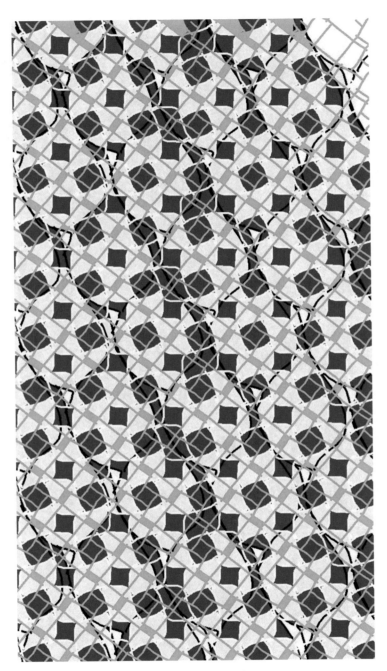

Fig. 1.21   Azure: Azure abstract shapes scattered in an intertwined mesh of ocean green

Fig. 1.22 Star-studded: Bright and radiant illumination in a star-studded night sky

Fig. 1.23 Sparkle: Rising sun sparkling and spreading light and, in the morning, sky dispelling the darkness

Fig. 1.24   Hues: Many-hued abstract shapes placed aptly giving the appearance of a floral design

Fig. 1.25 Scorching: Scorching bright rays in the background of a crisscross mesh of quadrangles with a cross-shaped corner

Fig. 1.26 Lattice: Abstract shapes with blue highlights scattered in a star-studded lattice of squares

Fig. 1.27   Variegated: Variegated intangible mesh emerging from a floral center

Fig. 1.28 Elliptical: A framework of elliptical shapes arranged on a polychrome plane

Fig. 1.29 Quatrefoil: Quatrefoil clover pattern highlighted in a vivid and colorful abstract background

Fig. 1.30 Embed: Octagon silhouette embedded in a four-sided border and intangible curves

Chapter 2
Arrows

Arrow, which can be seen as a prominent part of contemporary art, is one of the oldest and most basic graphics symbol. The discovery and use of arrow as a hunting weapon by the early human race give us an insight into the origin of human intelligence to find smarter ways to exist and endure. It is a tool symbolizing protection and sustenance of life. It symbolizes the skill and ability of the user to hit the target. It is also a symbol of direction to enable us to navigate through virtual and physical worlds.

Arrow in art is timeless. It is an idea converted into a symbol. In ancient arts, it is used as a symbol to represent ideas bigger than us. The interpretation of art with an arrow varies based on the direction and type of arrow portrayed. For instance, an arrow piercing a heart is used to symbolize tortured romanticism, and crossed arrows represent friendship and close ties.

In this chapter, designs with arrows as an integral part are presented. The vibrant colors and shapes in these designs will excite the visual sense of the viewer and will enliven their mood. Inspirations from ancient arts have been customized to create colorful and vibrant designs (Figs. 2.1, 2.2, 2.3, 2.4, 2.5, 2.6, 2.7, 2.8, 2.9, 2.10, 2.11, 2.12, 2.13, 2.14, 2.15, 2.16, 2.17, 2.18, 2.19, 2.20, 2.21, 2.22, 2.23, 2.24, 2.25, 2.26, 2.27, 2.28, 2.29 and 2.30).

K. Murugesh Babu et al., *Abstract Pattern Illustrations for Textile Printing*, Textile Science and Clothing Technology, https://doi.org/10.1007/978-981-16-5975-1_2

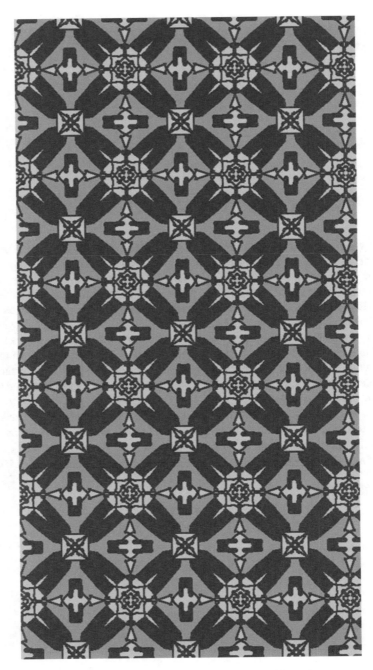

Fig. 2.1 Carefree: A blithe pattern placed in high spirits with repetitive pointers

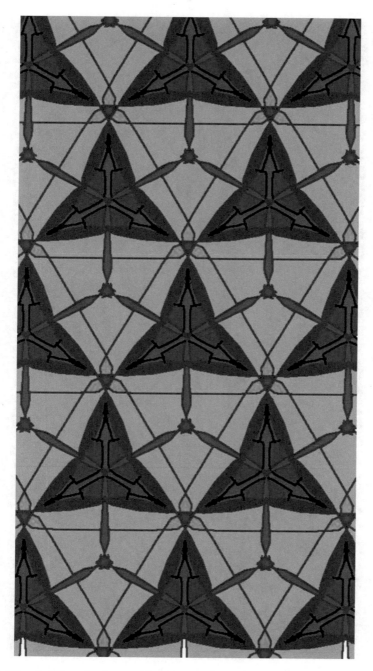

Fig. 2.2 Triangle: Triangular shapes with arrows creating an expression of apparent movement

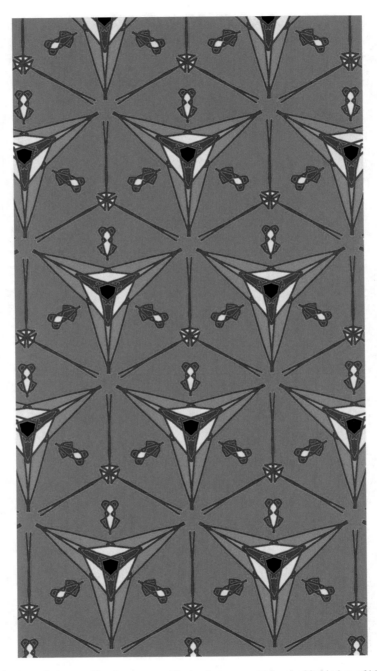

Fig. 2.3 Crystal: A crystallized structure with piercing arrows and embedded in hue of blue

Fig. 2.4 Pointer: An energetic prototype placed in a dynamic background with repeated pointers

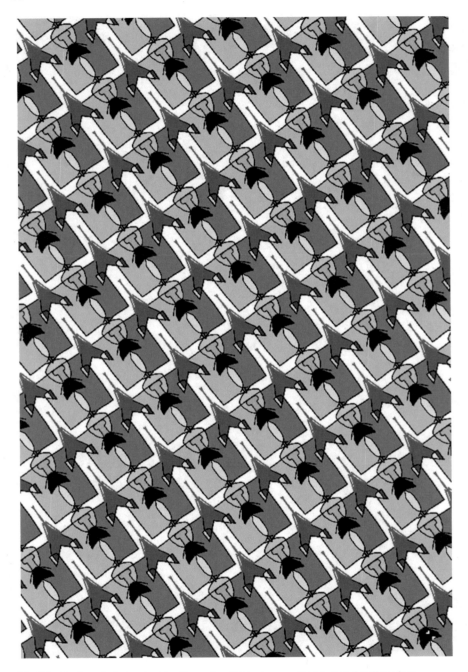

Fig. 2.5 Dartboard: A structured pattern of multiple arrows adhering to a dazzling plane

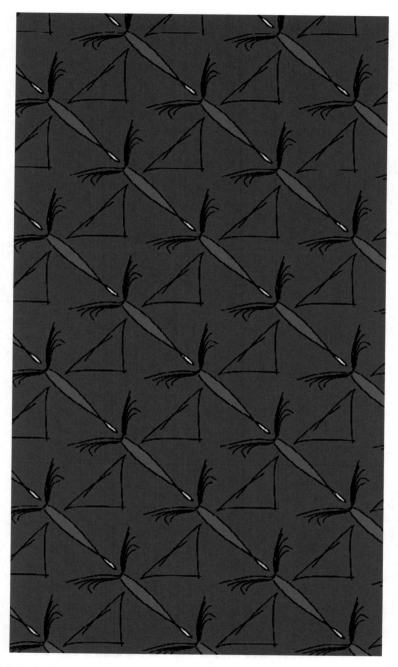

Fig. 2.6 Projectile: A structured framework of projectiles on a journey to infinity and beyond

Fig. 2.7 Conviviality: Cheerfully implanted fresh and curvy patterns bringing warmth and delight

Fig. 2.8 Vintage: Slender and focused arrows entrenched in an ordered scaffold of vintage pattern

Fig. 2.9 Winged: Sturdy and focused arrows with wings geared up for flight

Fig. 2.10 Guides: Multicolor segmented patterns with guiding arrows to route the directionless

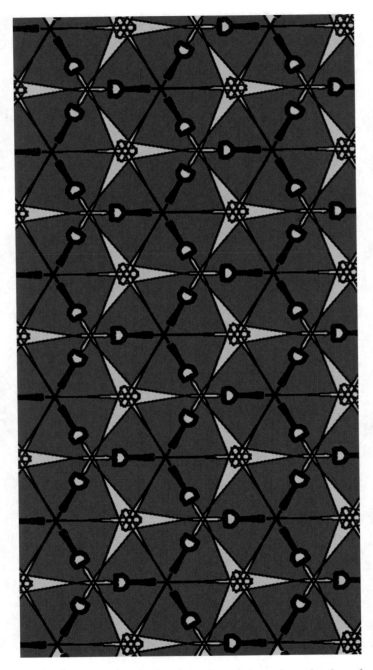

Fig. 2.11 Fortified: An intense fortification of triangles and arrows decorating the surface deck

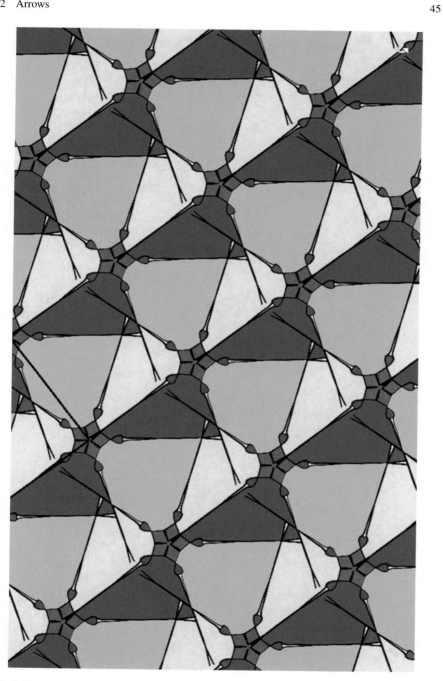

Fig. 2.12 Delineate: Abstract geometric shapes with a floral highlight outlined by arrows

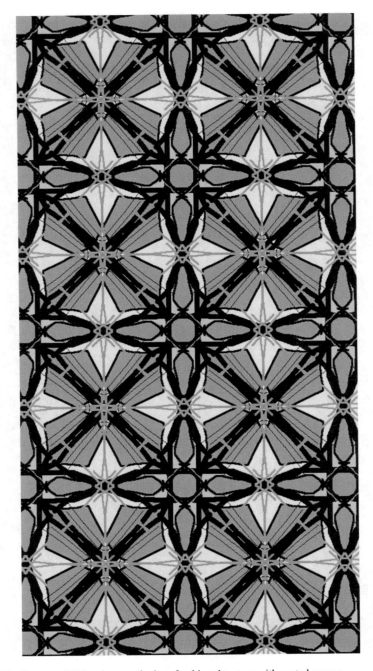

Fig. 2.13 Pleasant: Mild and warm shades of unbiased texture with rooted arrows

Fig. 2.14 White vine: The highlighted mediacs with apparently focused outlining

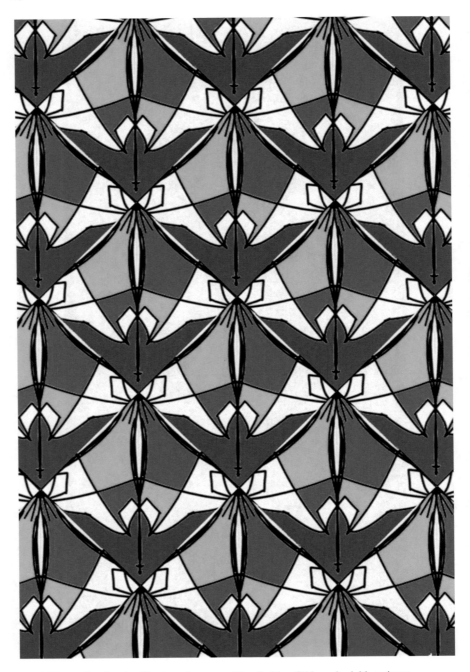

Fig. 2.15 Marine: Passive blue-toned pattern with a flashing fill in and prickly pointer

Fig. 2.16 Radiation: Energetic illusion of particles with emission into progressive pattern

Fig. 2.17 Trim: Recurring motif forming a binding and interactive surface appliqué

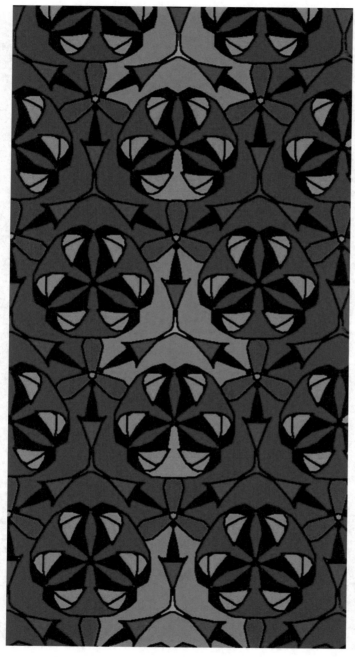

Fig. 2.18 Perennial: The mediac pattern created with continually recurring arrows and leaves

Fig. 2.19 Paragon: Pop mesh regarded with square repetition along with the quality appeal

Fig. 2.20 Synopsis: A focused and enhanced blueprint providing a summary of the highlights

Fig. 2.21 Prehistoric: A linear arrangement of prehistoric arrowheads in a bicolor plane

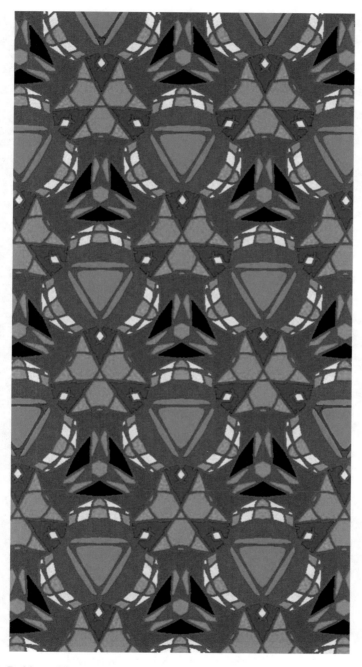

Fig. 2.22 Red-knox: Vintage geometric abstract with a simple green hexagon

Fig. 2.23 Silhouette: Olive green abstract sketch taking the silhouette of an arrow

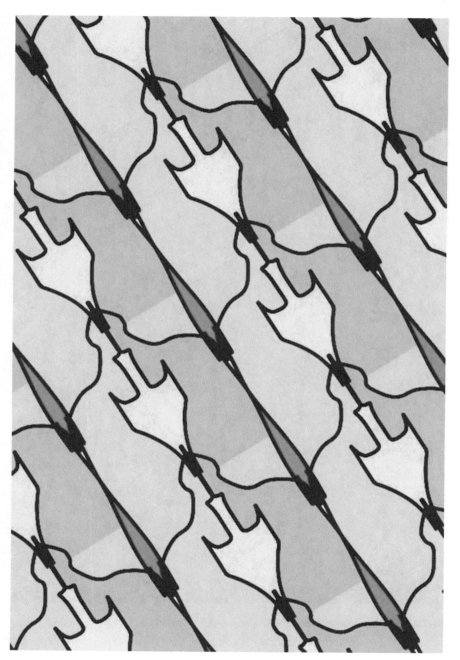

Fig. 2.24 Alternate: Yellow arrowheads alternated by blue pointers in a pale background with abstract curves

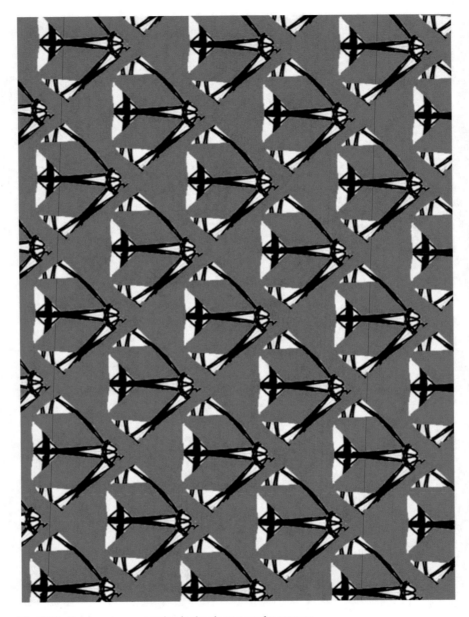

Fig. 2.25 Soaring: Arrows soaring in the air on a scarlet scenery

Fig. 2.26 Primitive: An assortment of primitive arrowheads decorated in a circular mounting

Fig. 2.27 Arrow mesh: A mesh of back-to-back connected arrows forming a quadrangle

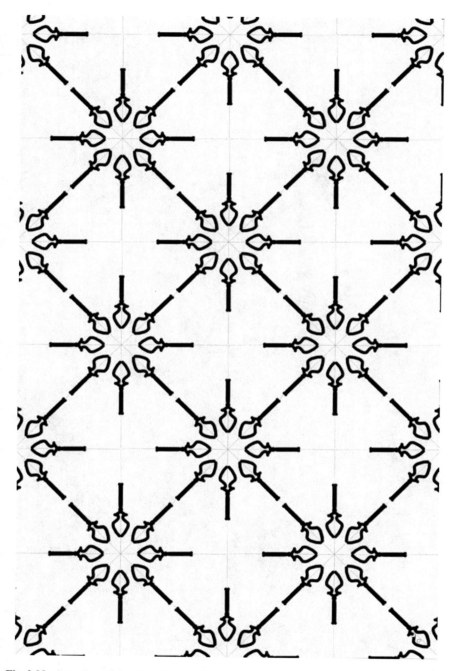

Fig. 2.28 Attractive: An outline of manifold arrows forming a simple and attractive pattern

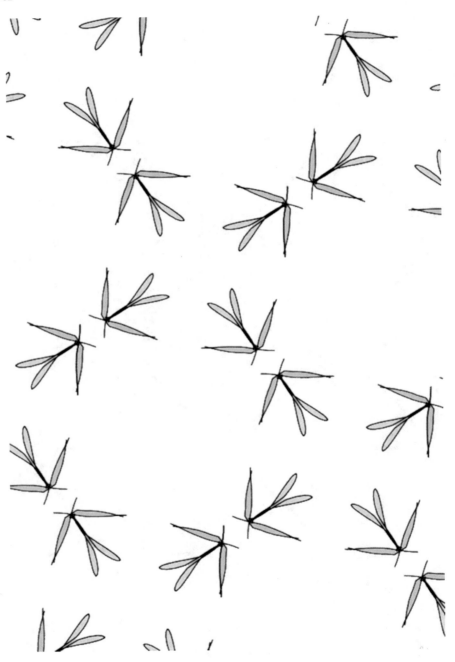

Fig. 2.29 Twosome: A pair of airborne arrows facing each other in a blank space

Fig. 2.30   Entrench: Abstract shapes with an embedding of arrows in shades of blue

Chapter 3
Arcs

An arc, mathematically defined as a part of a circle, can be used to depict any curved shape ranging from the bow shape of ballerina's arm to the elegant arch formed by a vine over a fence. The word 'arc' was used in the early centuries to describe the apparent path of a celestial body such as the sun over the horizon. An arc is a prominent shape seen in nature, from the halo of stars and the outline of a rainbow to the contour of a natural cliff. It is a shape with divine inspiration and inspires the observer to think beyond the limits.

Arc shapes have been an imperative part of historical arts and designs and continue to be an indispensable part of the contemporary art world. Designs with an iconographic arrangement of arc shapes must be an essential part of a designer's palette to create awe-inspiring designs. Designs with arcs go beyond cultural and artistic boundaries that open up our minds to endless imaginations and interpretations.

In this chapter, designs with a variety of arc shapes, textures and colors are presented. The lively colors and shades in these designs will inspire the designers to put on display their creativity and provide them with an excellent opportunity to contemplate designs on fabrics and textiles as a form of artistic expression (Figs. 3.1, 3.2, 3.3, 3.4, 3.5, 3.6, 3.7, 3.8, 3.9, 3.10, 3.11, 3.12, 3.13, 3.14, 3.15, 3.16, 3.17, 3.18, 3.19 and 3.20).

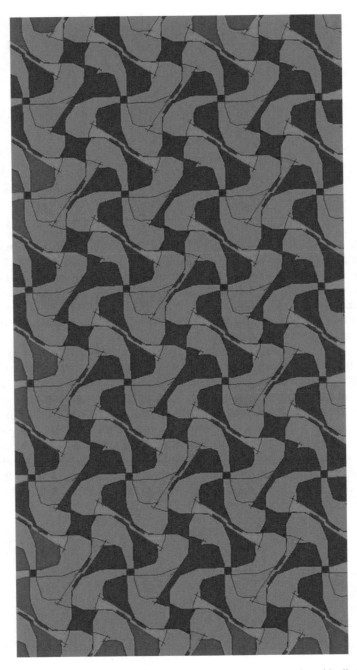

Fig. 3.1 Notion: Beautiful modern abstract having a delicate and warm color with all elements of motif characterized by exquisite

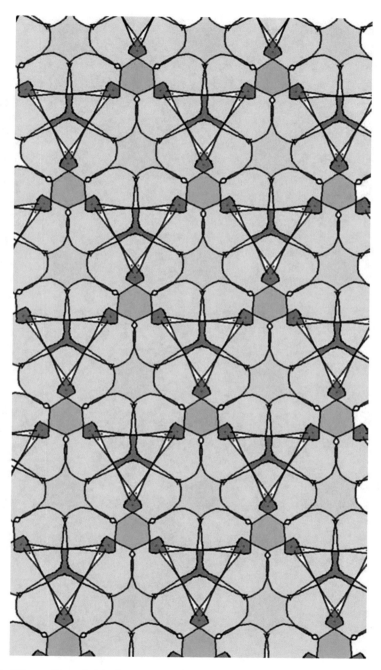

Fig. 3.2 Filigree: Recurring lattice patterns with bold elements in neutral hues and stroked by cross-black

Fig. 3.3 Fortress: Arcs and triangles projecting part of a rampart from an angle with equal directions

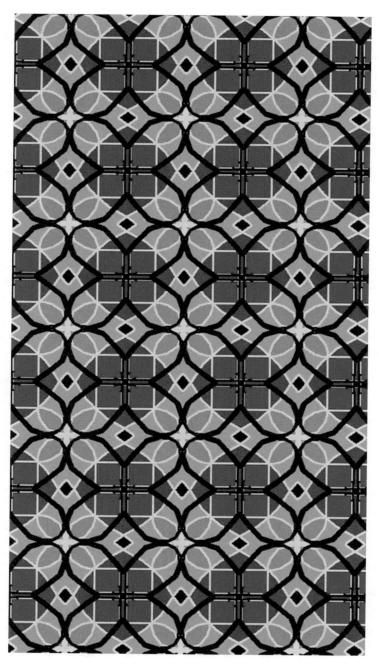

Fig. 3.4 Geniality: Endowed in a cheerful manner, the hues create fresh and curvy floral vine

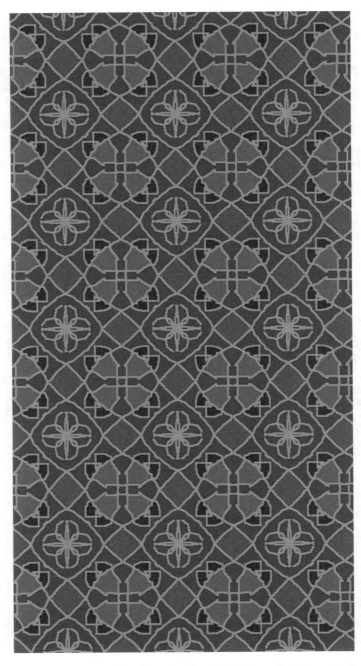

Fig. 3.5 Efflorescent: Migrated clover loops forming a blend of two floral shapes on the surface

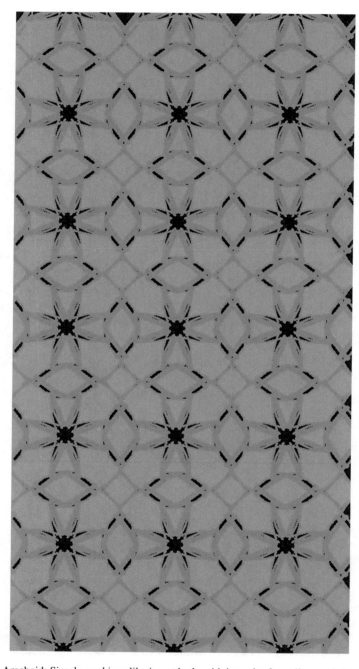

Fig. 3.6 Arachnid: Simple markings like insect body with irregular focus lines and shadows

Fig. 3.7 Keystone: A neutral center at the summit of joining and locking the other whole shapes together

Fig. 3.8 Immaculate: Seamless abstract pattern with the color crush

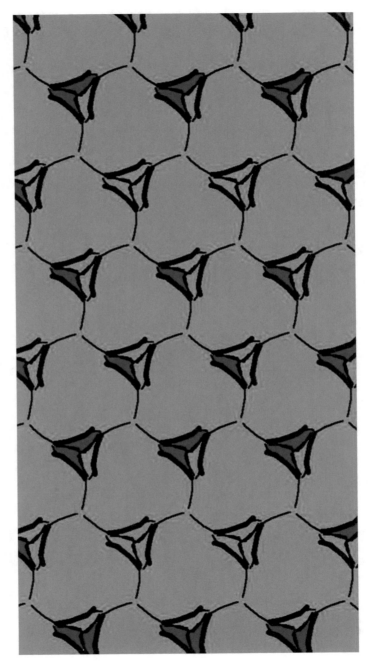

Fig. 3.9 Sporadic: Classic blue design featuring irregular patterns of hexagons that are true expression of nature similar to honeycomb

Fig. 3.10 Decanter: Decorative mesh with beautifully placed three-sided shapes intersecting with crisscrosses

Fig. 3.11 Flabellum: Triangular fan-shaped repeated motifs with warm color insertion at irregular intervals

Fig. 3.12 Agitate: Colored shadows in contrary color flustering the briskly distributed shapes into
an abstract pattern

Fig. 3.13 Oriel: simple pattern curated delicately using spherical shapes in a minimalistic way

Fig. 3.14 Charter: The supporting and subtle persistent shapes framing a structured pattern

Fig. 3.15 Floweret: Double tone repetitive botany cluttered closely making a tight flower head of a composite

Fig. 3.16 Dendritic: Abstract pattern is done with branch form shapes resembling darker dendrites of a tree

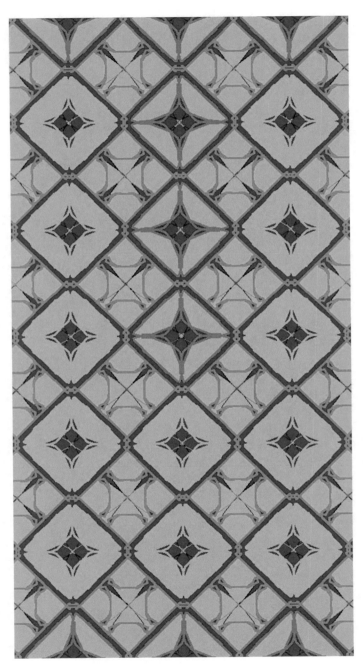

Fig. 3.17 Luminary: An influencing pattern with prominent rhombus and highlights in blue

Fig. 3.18 Tranquility: Peaceful and calm amalgamation of shapes giving a sense of sunset and shade together

Fig. 3.19 Sophisticated: Relaxed linear deal of florals developing a non-complex pattern

Fig. 3.20 Layoff: Striking pattern in blues downsizing with the repetition of leaf motifs

Chapter 4
Chevrons

A chevron is a V-shaped design with two diagonal stripes meeting at an angle and is seen usually on the sleeve of a uniform to indicate rank and service. Different shapes and symbols carry a lot of thoughts and beliefs from the people of the past, and a chevron shape has seen a sign of heraldry. It has been in use from early days on ceramics, pottery and rock carvings. The word chevron has a French origin and is derived from its resemblance to the pattern of building rafters.

Designs with chevrons help the designer to create attractive patterns with timeless appeal. The symmetrical patterns with chevrons result in beautiful shapes and designs and have been the foundation for fashion in textile designing.

In this chapter, a variety of designs with diverse chevron patterns and vivid colors have been presented. The patterns and details in the designs will enable the designers to create unique designs and provide them with endless design possibilities (Figs. 4.1, 4.2, 4.3, 4.4, 4.5, 4.6, 4.7, 4.8, 4.9, 4.10, 4.11, 4.12, 4.13, 4.14, 4.15, 4.16, 4.17, 4.18, 4.19 and 4.20).

© The Author(s), under exclusive license to Springer Nature Singapore Pte Ltd. 2022   87
K. Murugesh Babu et al., *Abstract Pattern Illustrations for Textile Printing*, Textile Science and Clothing Technology, https://doi.org/10.1007/978-981-16-5975-1_4

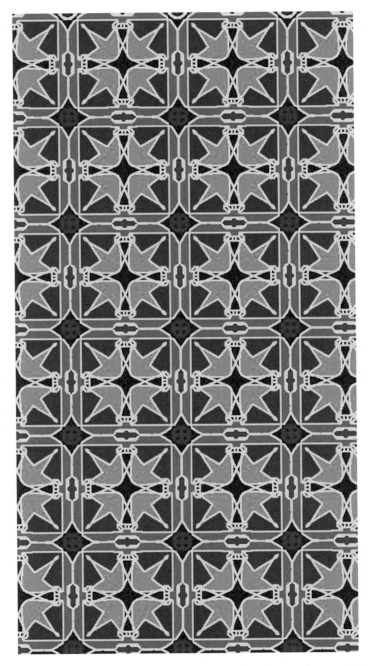

Fig. 4.1 Ultramarine: Grinded shades of blue spread across like pigmented geometrical shapes

Fig. 4.2 Ambition: Strong and focused arrowed rectangles turning into highlighted chevrons in the middle

Fig. 4.3 Headliner: Flying chevrons in contrary hues depicting motion illusion

Fig. 4.4 Paramount: The intrusting and intense out bloom with the supreme importance of interlacing shapes

Fig. 4.5 Sovereign: Royal and powerful motif minted in commemorative pattern resembles supremacy

Fig. 4.6 Salvage: Olive wrecked shapes salvaged into a pattern under way

Fig. 4.7 Petiole: Slender stalks like joining leaf to stems creating two different structures visually

Fig. 4.8 Cuneate: Iconic arc yellow triangle with an infinite number of green display art on a black background

Fig. 4.9 Ninja-star: Traditional pattern that displays the contour of sharp throwing star blade

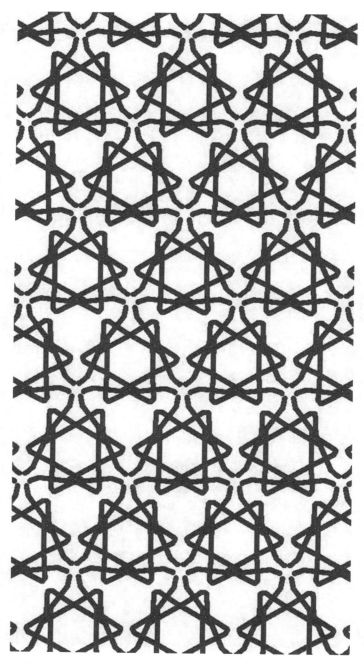

Fig. 4.10 Coherent: Interference pattern of seamlessly interlocking star and triangle structures

Fig. 4.11 Secluded: Vintage-inspired mandala decoration design isolated on royal blue flourishing the pattern illuminating movement

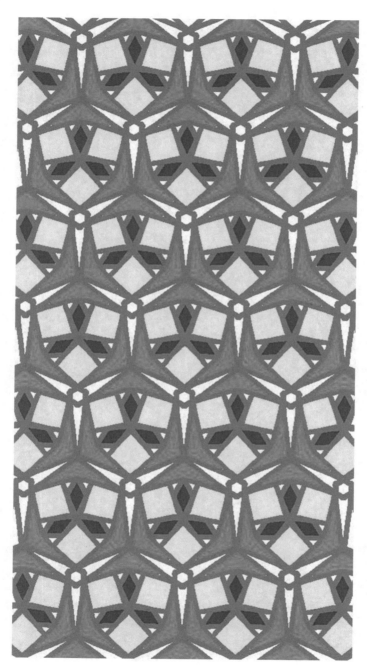

Fig. 4.12 Bastion: Lines and hexagons projecting part of a bastion from an angle with equal directions

Fig. 4.13 Shrubbery: Equal sized perennial inverted and regular triangles showing negative and positive spaces in this abstract focused pattern

Fig. 4.14 Cincture: Spherical ring structures with pinked edges in cool colors and pop detailing making it a bohemian abstract

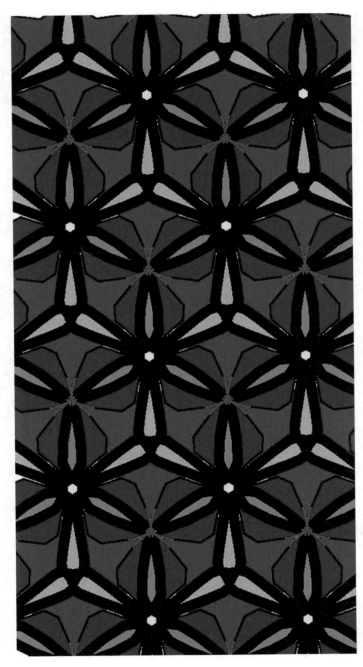

Fig. 4.15 Intermingle: Three edged shapes mingled together with measured space between producing a very clean pattern

Fig. 4.16 Luminary: An influencing pattern with prominent stars in blue Background

Fig. 4.17 Brilliant: Skillfully used shaped mortifying into well-acted floral mesh

Fig. 4.18 Impression: Effect produced with graphic evidence of an entertaining action on surface

Fig. 4.19 Transformer: One triangle transforming into a hexagon and vice versa to set up a signaled pattern of high quality

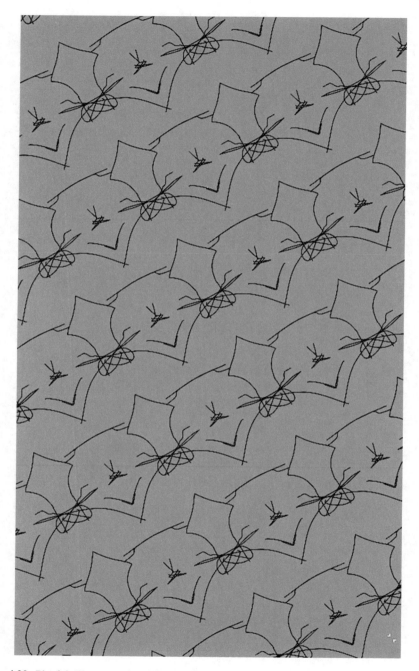

Fig. 4.20 Playful: The amused and light-hearted motif in warm hues gives a soothing punch of movement within the surface texture

Chapter 5
Floral

Flowers speak in a language of their own with each variety and color of blossom whispering its secrets. Each flower has its own special meaning and symbolizes something very different from the other. Some flowers tell us the story of love and devotion while some revitalize us during our time of sickness. Every flower color from pale blue to bright red offers a profound and significant story. The flower color is as important in the arrangement as the flower type.

For centuries real flowers were used in the form of wreaths and brooches to adorn outfits and add a sweet-scented touch to the attire. From ancient days, flower designs have also been used to create beautiful patterns for decoration and fabric prints. In today's world, floral designs have become an indispensable part of our designs and art. The cheerfulness and positive energy that floral designs bring into our life are unparalleled.

In this chapter, floral designs with meaningful use of colors have been presented. The profound symbolic value portrayed by these floral designs is sure to capture the imaginations of the viewer vividly. Attires with such floral designs will not only be beautiful to look at but it will also brighten the outfit (Figs. 5.1, 5.2, 5.3, 5.4, 5.5, 5.6, 5.7, 5.8, 5.9, 5.10, 5.11, 5.12, 5.13, 5.14, 5.15, 5.16, 5.17, 5.18, 5.19, 5.20, 5.21, 5.22, 5.23, 5.24, 5.25, 5.26, 5.27, 5.28, 5.29 and 5.30).

K. Murugesh Babu et al., *Abstract Pattern Illustrations for Textile Printing*, Textile Science and Clothing Technology, https://doi.org/10.1007/978-981-16-5975-1_5

Fig. 5.1 Outbloom: Exceeding floral pattern with appealing and intense out bloom

Fig. 5.2 Castor: Swiveling wheels joining kinked horns turns into a floral pattern

Fig. 5.3 Herbaceous: Floral structure denoting non-stem flora and creating lively growth

Fig. 5.4 Supreme: An influencing pattern with prominent rhombus petals and purple highlights

Fig. 5.5 Warped: Warped floral shapes with a self-designed sturdy pattern

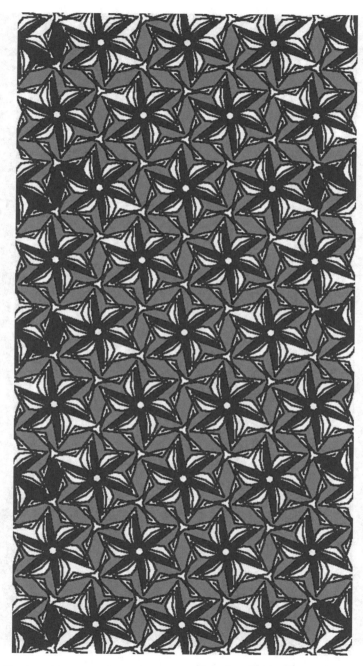

Fig. 5.6 Narcissus: Dense abstract pattern with a compact structure of blue and green narcissus flower

Fig. 5.7 Floral mesh: Proficiently used shape turning into well-acted floral mesh

Fig. 5.8   Roseate: Early pink light parting with lavender plumage of Florina

Fig. 5.9 Jalousie: Vibrant rows of floral crepes creating a pattern of bright angled slats

Fig. 5.10 Sylvan: Association of pastoral graphic and floral woods like a rural expression

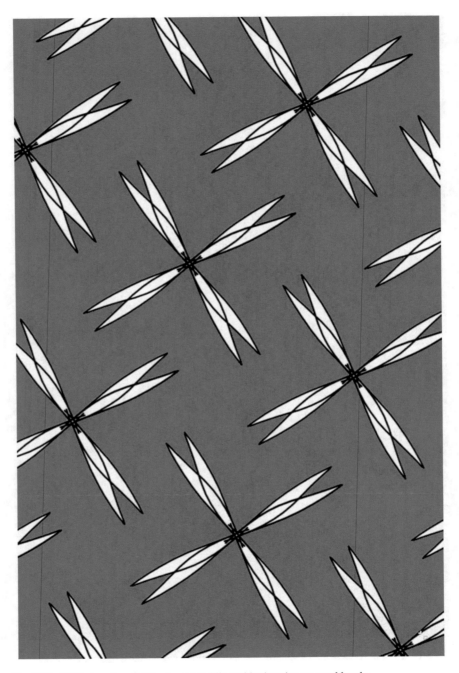

Fig. 5.11   White jasmine: Brilliant and sparkling white jasmine on a red locale

Fig. 5.12 Graceful: Medieval floral design in a light background that is charmingly and pleasantly simple and graceful

Fig. 5.13 Bay window: Polygonal shaped floral pattern creating an expression of outward movement in an order

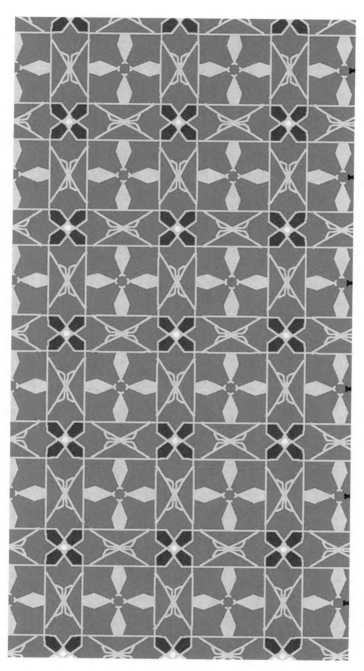

Fig. 5.14 Lively: Florals creating a delusion of a pleasant and bright pattern like a gentle wind

Fig. 5.15 Charming: Simple yet charming floral outline with serenity and enchanting abstract

Fig. 5.16 Botanic: Flora life creating an aqua survey through color notes

Fig. 5.17 Montage: A colorful polar extending the cool impression into floristic abstract

Fig. 5.18 Clematis: Elegant dark blue densely packed clematis flower pattern

Fig. 5.19 Seamless: Medieval seamless pattern on a bright multicolor background with elaborate and decorative floral motifs

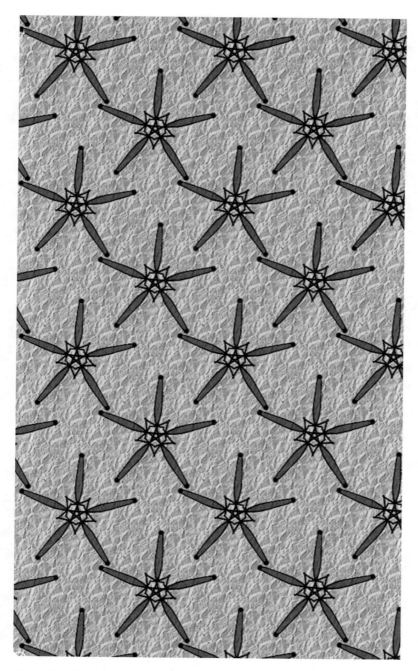

Fig. 5.20 Serene: Floral design that is nice-looking and serene with captivating backdrop

Fig. 5.21 Horticulture: Art of lightning curves gardening the texture like looped in crosses

Fig. 5.22 Gyration: A whirling motion created by circles and cones appears as rapid movement on the surface

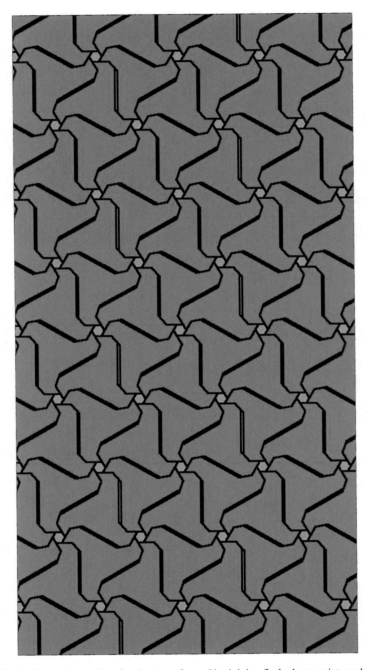

Fig. 5.23 Outbloom: Exceeding floral pattern formed by joining flashed one point garden

Fig. 5.24 Tranquillity: Peaceful and calm amalgamation of shapes giving a sense of sunset and shade together

Fig. 5.25 Intense: An arrangement of intense yellow floral patterns with brunette highlights

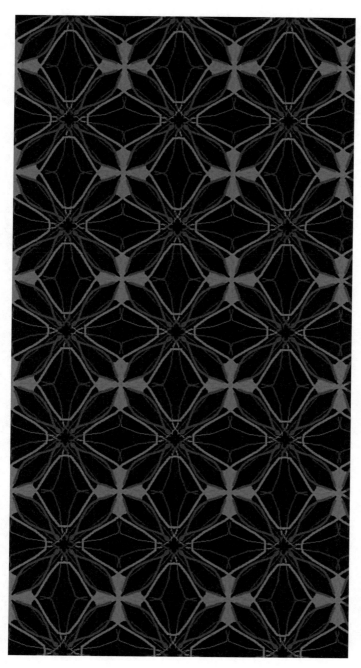

Fig. 5.26 Structure: A conscious thought in light producing a pictorial imitation

Fig. 5.27 Desistance: Focused action on edges of the cube with focused hue and shape

Fig. 5.28 Bulwark: An extended rampart of triangles and pentagons as linings above the surface deck

Fig. 5.29 Yellow rose: An optimistic pattern using emphasized rose as repeated shape creating a verb of abstract print

Fig. 5.30 Fragment: A large shape slashed off evenly with the fine contour in brick red

Chapter 6
Circles and Curves

A curve is a simple shape that connects two points with bends and dips and turns. At the same time, it is an extremely unpredictable shape that can take different thicknesses, orientations and curvatures. A circle is one of the most basic curved shapes. The emotion linked with the curved line varies from serene and simple to vigorous and vivid based on the variance of shape. Connections and intersections of curves with different orientations result in patterns with purposeful groupings.

Curves that are used along with various other design elements are interpreted based on the elements around them. The fluid and soft nature of the curves can increase visual interest and attach emotions to a design. Open and closed circular curves can emphasize one component in the design over others. Curves, though a simple design element made of a single stroke, can have a remarkably impressive impact when added to a design.

In this chapter, colorful and insightful designs with circles and curves are presented. The textures, connections and placement of curves in the design form unique patterns. The role performed by the circles and curves in the overall design scheme is incomprehensible and would serve as a platform for designers to create dynamic patterns (Figs. 6.1, 6.2, 6.3, 6.4, 6.5, 6.6, 6.7, 6.8, 6.9, 6.10, 6.11, 6.12, 6.13, 6.14, 6.15, 6.16, 6.17, 6.18, 6.19, 6.20, 6.21, 6.22, 6.23, 6.24, 6.25, 6.26, 6.27, 6.28, 6.29 and 6.30).

Fig. 6.1 Rings: A framework of sea-green interacting rings on a white background

Fig. 6.2 Pennon: A round of delicate feathers imitating a floral pattern at irregular intervals in cool and peaceful hues

Fig. 6.3 Intertwined rings: A framework of intertwined rings of orange and gray

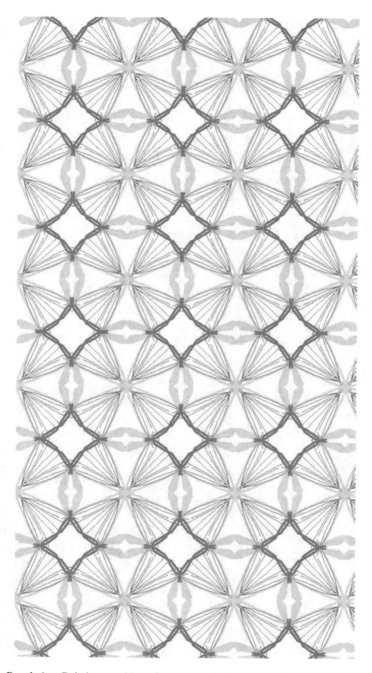

Fig. 6.4 Revolution: Relative repetition of a motion and shape perceiving the textural prowess

Fig. 6.5 Vintage: Vintage inspired decorative rings with colorful extensions

Fig. 6.6 Concentric: A simple yet powerful arrangement of concentric rings bordered with abstract patterns

Fig. 6.7 Mandala art: Circular simple black and white mandala representing deep thinking, mystery and individuality

Fig. 6.8 Heart pie: Heart shape on the edge looks metaphorical and symbolizes as a centered pattern with emotions and love

Fig. 6.9 Blossom: White multilayer blossoms in a yellow background

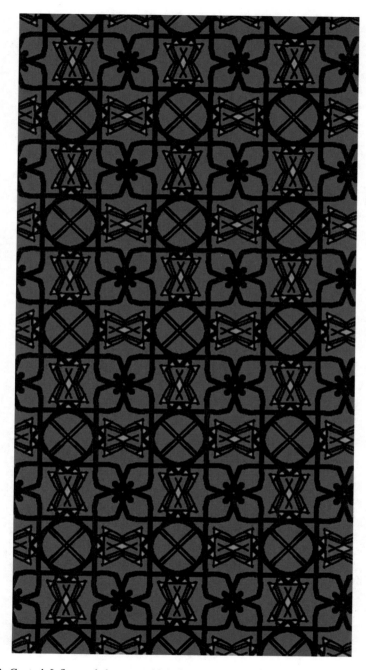

Fig. 6.10 Control: Influenced chevrons with ball patterns complimenting in continuation

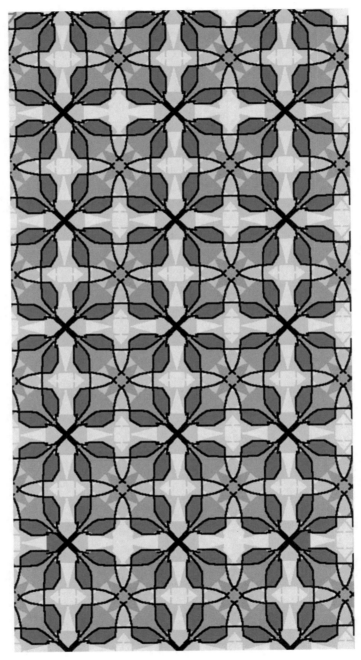

Fig. 6.11   Octagon: Circles divide from edges turning into a square with a regular leafy formations

Fig. 6.12   Ripple: Series of waves giving an impression of a water surface caused by a slight breeze

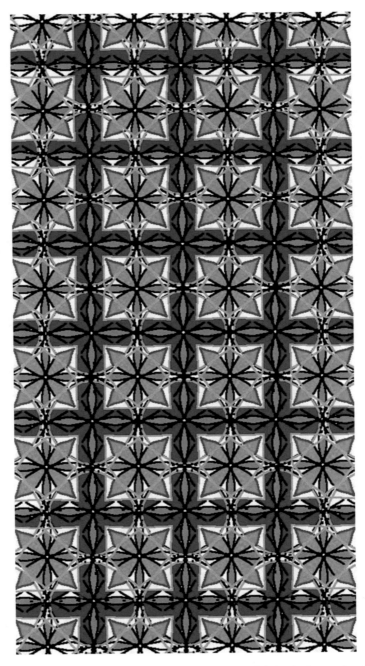

Fig. 6.13 Globule: Small square patterns of a substance against a brighter background and luminous center

Fig. 6.14 Bygone: Medieval seamless pattern on a bright multicolor background with elaborate and decorative floral motifs

Fig. 6.15 Posy: Bunch of scattered flowers in line like a scribble attempting to impress the cool background

Fig. 6.16 Intricate: Intricate pattern curated skillfully using indigo spherical shapes

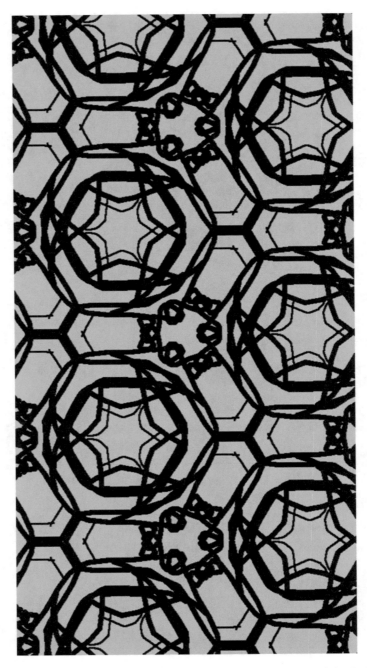

Fig. 6.17 Hoary: Cobwebs of shapes, star centric moving pattern with repeated circular lines

Fig. 6.18 Hazardous: Bright pattern creating an illusion of circular molds but advertising by the use of the color black

Fig. 6.19 Clement: Color codes with a gentle blend creating warmth and balanced texture

Fig. 6.20 Blueprint: Fine line photographic framework with sensitivity and white grounds

Fig. 6.21 Coarse: Coarse edged egg-shaped patterns strewn on a white surface

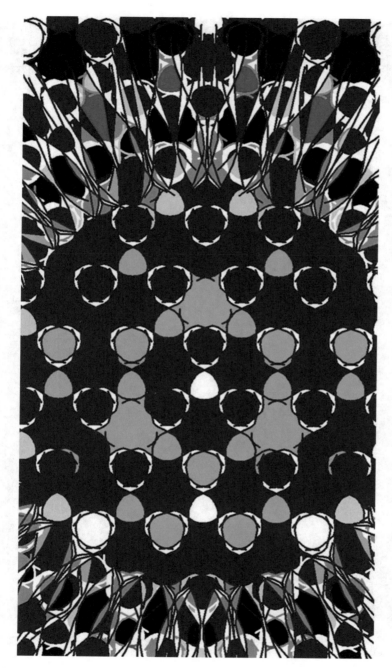

Fig. 6.22 Eyes: Piercing eyes peeping out from the dark and gloomy woods

Fig. 6.23 Harmonize: Elliptical rings surrounded by twofold border complementing the background tint

Fig. 6.24 Three dimensions: Three-dimensional prototype approaching an umbrella with vibrant and energetic shades

Fig. 6.25 Illusion: Multicolored and multifaceted concentric rings giving an illusion of outward movement

Fig. 6.26 White daisy: White daisies radiating on a crimson surface

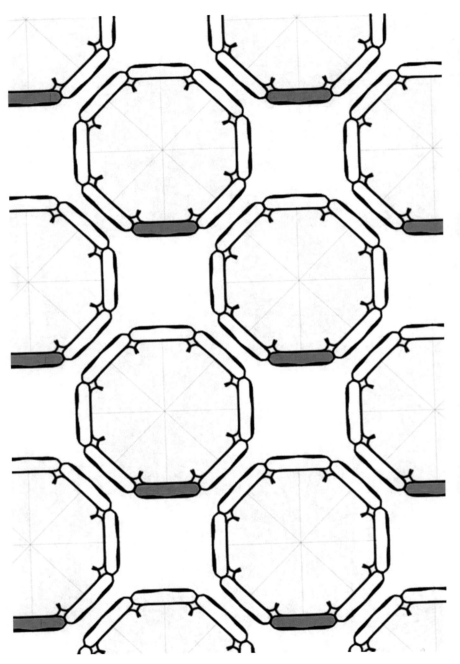

Fig. 6.27 Circular frame: A structured pattern of octagons extending to a circular frame

Fig. 6.28   Floral rings: Decorative floral rings in a framework of abstract patterns

Fig. 6.29 Wheels of life: A framework of blue circular patterns depicting wheels of life

Fig. 6.30 Banner: An encompass of elusive rings emulating a flowery outline in cool colors

Chapter 7
Clover Loop

A clover loop is a quatrefoil structure of partially overlapping circles. It is a symmetrical decorative design element found in art, architecture and traditional symbolism. The clover shape is derived from four-leafed clover but, in general, applies to four-lobed shapes in diverse contexts. A similar shape with three lobes is a trefoil. Clover shapes are used in artwork or decoration incorporating decorative or ornamental interlacing or knotwork.

Through centuries, clover shapes have been regarded as Celtic charms. A three-leaf white clover stands as Ireland's national symbol and is considered as sacred. A clover shape denotes the feeling of faith, hope, love and luck and is often worn by couples due to tie knots.

For centuries, clover loops have been used in arts and designs around the world, and designers and artists still use it today, though it has been reinterpreted and recontextualized based on its use and purpose. It is a timeless design element that continues to fascinate the artists and add charm and meaning to the artwork.

In this chapter, designs with clover loops have been presented. It provides the designer with a palette of inexhaustible array of dynamic patterns to create designer fabrics (Figs. 7.1, 7.2, 7.3, 7.4, 7.5, 7.6, 7.7, 7.8, 7.9, 7.10, 7.11, 7.12, 7.13, 7.14, 7.15, 7.16, 7.17, 7.18, 7.19 and 7.20).

K. Murugesh Babu et al., *Abstract Pattern Illustrations for Textile Printing*, Textile Science and Clothing Technology, https://doi.org/10.1007/978-981-16-5975-1_7

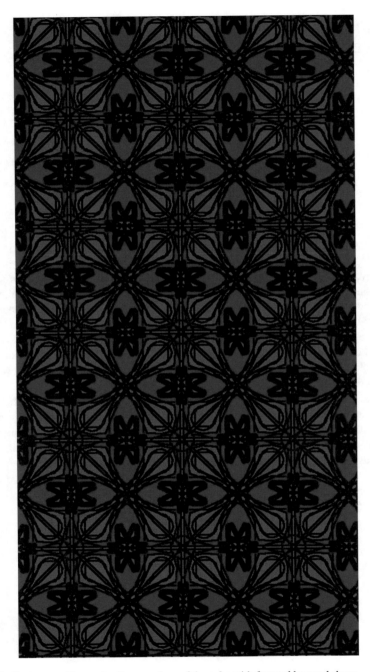

Fig. 7.1   Desistance: Focused action on edges of the cube with focused hue and shape

Fig. 7.2 Classic: Fine tribal art with a classic mix of colors in different abstract shapes

Fig. 7.3 Tribal: Tribal artisanal multicolor floral motif in graphite blackbackground with a brilliant and sparkling white abstract embedded in between flowers

Fig. 7.4 Ethereal: The contrary colors in extremely delicate and light in a way that seems exquisite

Fig. 7.5 Perianth: Floral pattern emphasizing the outer sepals and petals of a flower enveloping into a structured surface

Fig. 7.6 Epitomized: Archaic distribution of shapes well placed representing floral pattern with geometrical space

Fig. 7.7 Moss green: Moss green clover leaf pattern in a graphite black background with flower embeddings

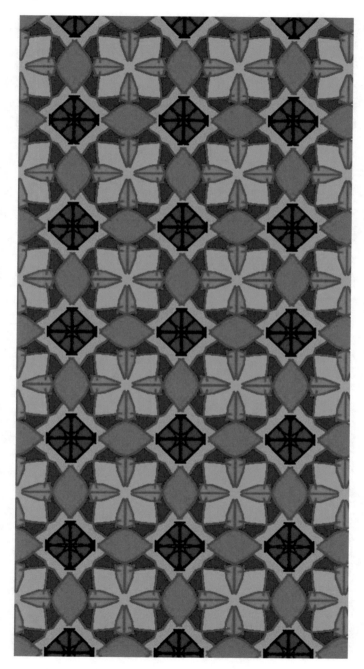

Fig. 7.8 Efflorescent: Migrated clover loops forming a blend of two floral shapes on the surface

Fig. 7.9 Pleasant: The expressions created by shapes and primary colors give a sense of agreeable contentment

Fig. 7.10 Variegation: The warm appearance stripping out of the expression making the pattern full of life

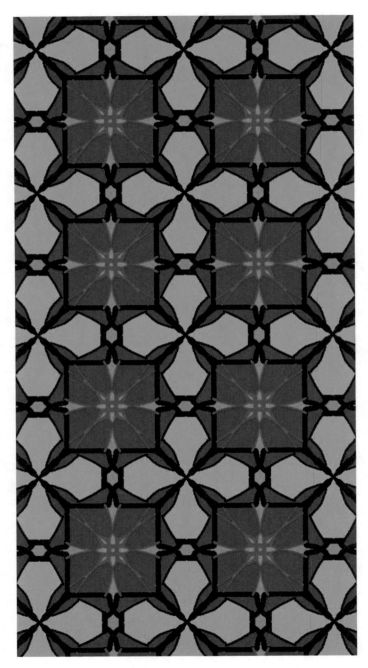

Fig. 7.11 Cease: Shapes used repeatedly into the geometrical floral pattern with white rhombus used as focus shaped pattern

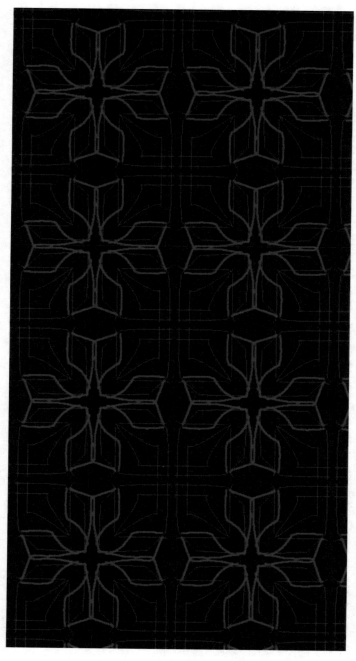

Fig. 7.12 Charming: Simple yet captivating floral pattern with stillness and Charm

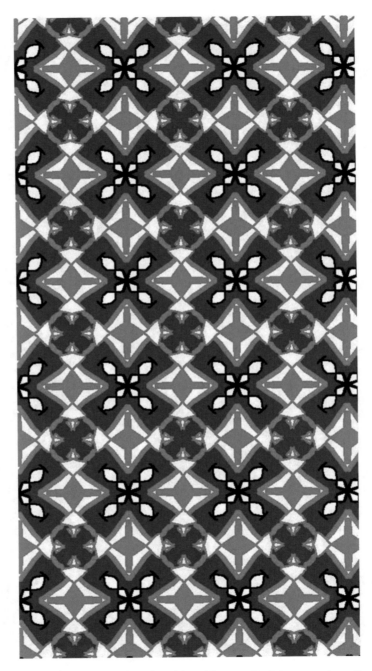

Fig. 7.13 Energy: Energetic delusion of particles with emanation into progressive pattern

Fig. 7.14 Aqueous: Aqueous flora life with color annotations

Fig. 7.15 Rustic: Association of rustic graphic and floral woods like a pastoral Expression

Fig. 7.16 Epitomized: Archaic distribution of shapes well placed representing floral pattern with geometrical space

Fig. 7.17 Vadalum: Floral design in a gloomy background that is charmingly and pleasantly simple

Fig. 7.18 Cherry: Clover shaped motif in cherry red background with abstracts embedded in between flowers

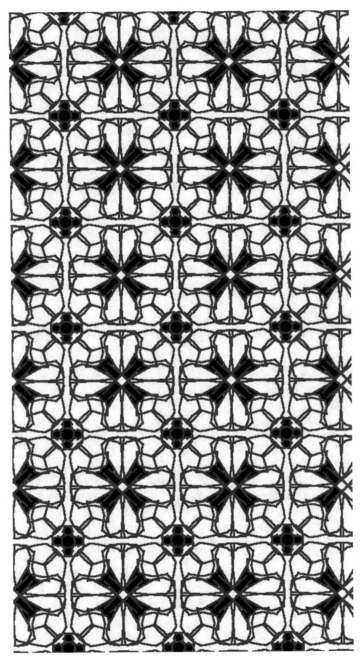

Fig. 7.19   Abridgment: A focused and enhanced design providing an outline of the highlights

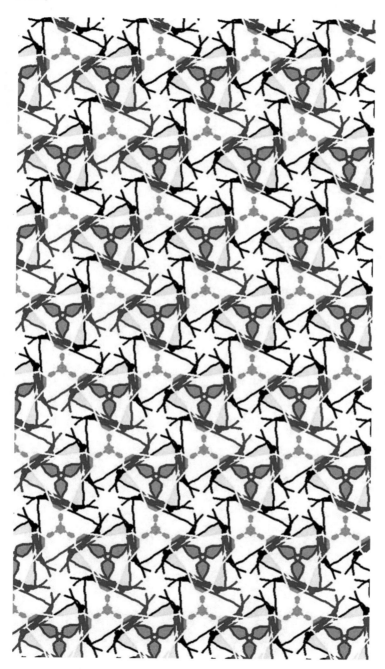

Fig. 7.20 Horticulture: Art of lightning curves gardening the texture like looped in crosses

Chapter 8
Criss-Cross Curves

A criss-cross design or pattern consists of a network of lines crossing each other. Criss-cross lines add a lot of value to a design and change the outlook of the design. Criss-cross patterns in designs help develop a mesh of abstract and other expressive shapes.

Criss-cross lines help to bring out the light and shadow in different parts of the design by varying the density of the lines. It helps the designers to develop designs that create an illusion of outward or inward movement thus creating a ripple or wave effect. Criss-cross lines also help to add multiple dimensionalities to the art thus making it more authentic.

Criss-cross patterns create cool artistic designs that radiate positivity and aid in creating a sense of gratification. This chapter presents an assortment of designs with criss-cross patterns embedded with other abstract and meaningful shapes. These patterns will serve as a guideline for designers to create beautiful and vibrant designs (Figs. 8.1, 8.2, 8.3, 8.4, 8.5, 8.6, 8.7, 8.8, 8.9, 8.10, 8.11, 8.12, 8.13, 8.14, 8.15, 8.16, 8.17, 8.18, 8.19 and 8.20).

K. Murugesh Babu et al., *Abstract Pattern Illustrations for Textile Printing*, Textile Science and Clothing Technology, https://doi.org/10.1007/978-981-16-5975-1_8

Fig. 8.1 Pivot: Diamond brackets toward the central point with fractured mesh as top layer

Fig. 8.2   Swoosh: The intensive move of the design representing a flash or strip of color

Fig. 8.3 Luminary: An influencing pattern with prominent rhombus and highlights in blue

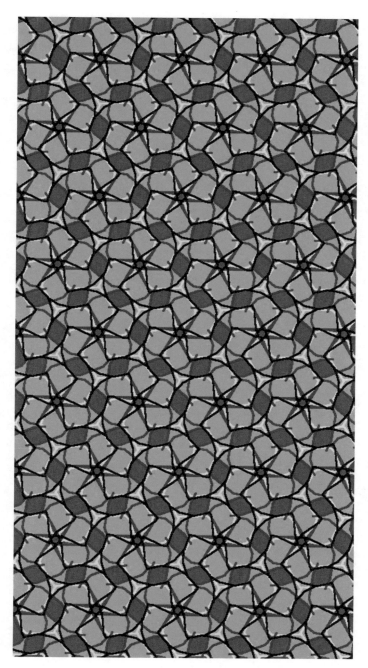

Fig. 8.4 Bloom: Double tone repetitive botany tangled closely making a constricted flower head of a composite

Fig. 8.5 Dendritic: Abstract pattern is done with branch form shapes like murkier dendrites of a tree

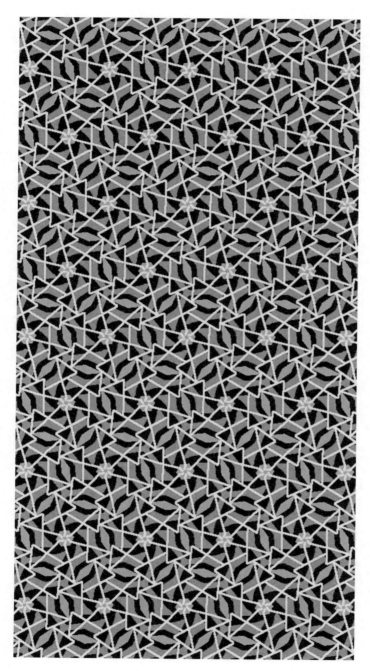

Fig. 8.6 Cobweb: A tangled three-dimensional dusty intricate spider mesh in warm colors

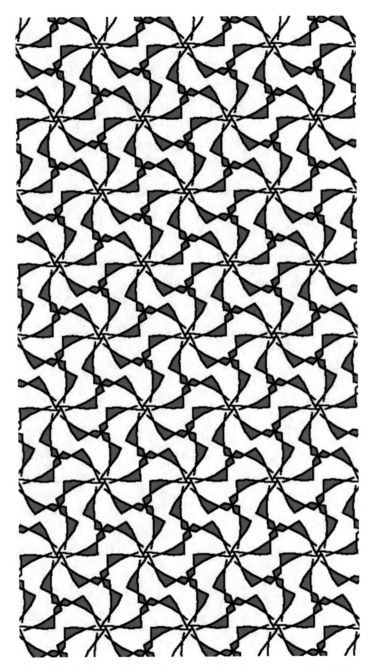

Fig. 8.7 Swivel: Mix of white and blue colors representing three sharp rotating blades

Fig. 8.8 Dovetail: Abstract design pattern with simple and traditional intermeshed shapes

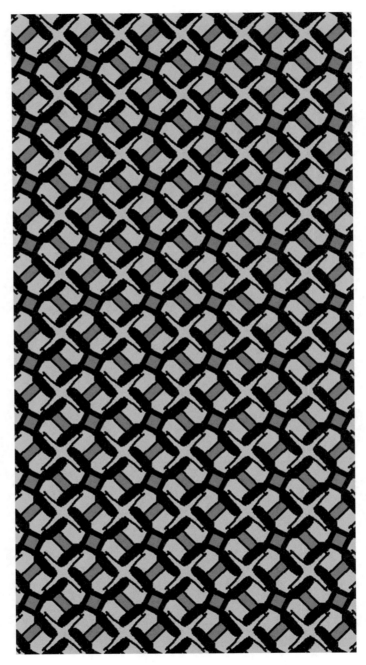

Fig. 8.9 Lattice: Repeated lattice patterns with valiant elements in neutral hues and patted by cross-black

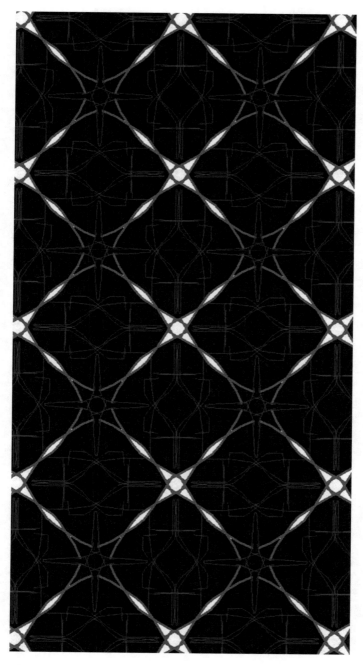

Fig. 8.10 Midnight: The transition moment of contrary color code creating equidistant peace pattern

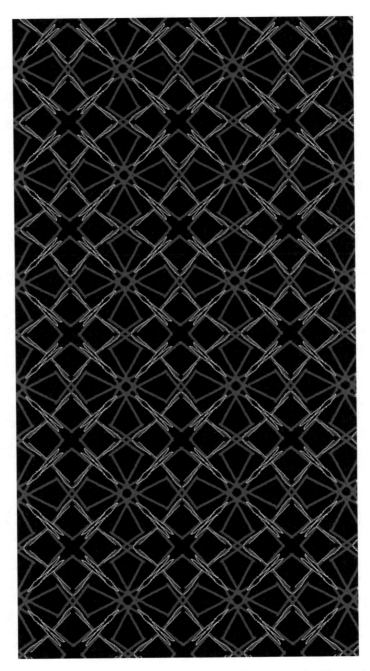

Fig. 8.11 Nightsest: Amalgamation of minimal shapes creating a nocturnal and impressive pattern in blue

Fig. 8.12 Button: Square protuberance on the surface marked by cross edges for a prominent pattern

Fig. 8.13 Fireworks: Diffusion of complimentary scheme entertaining the surface aesthetics

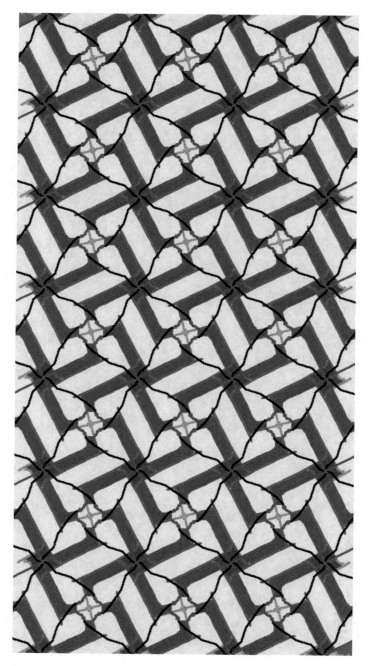

Fig. 8.14 Criss-crosses: A framework of criss-cross patterns on a yellow background with floral centers

Fig. 8.15   Jalousie: Vibrant rows of floral crepes creating a pattern of bright angled slats

Fig. 8.16 Splash: An intricate mesh of white and blue lines with random splashes of blue

Fig. 8.17 Formation: A lively manifestation created with structural base and contrast color scheme

Fig. 8.18 Fragment: A large figure cut off symmetrically with fine line intersections in brick red with blue crosses

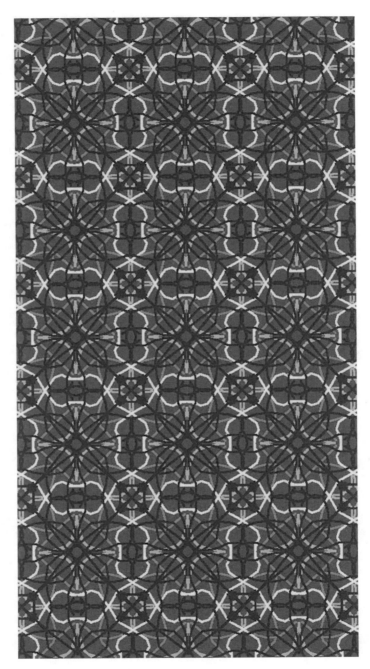

Fig. 8.19 Floral mesh: A beautifully woven floral mesh of multiple layers of multicolor on a green surface

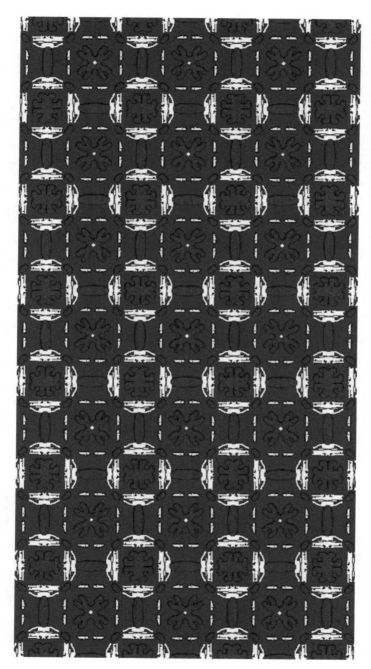

Fig. 8.20 Intricate: Intricate pattern curated delicately using floral shapes in minimalistic way

Chapter 9
Dreidels

A dreidel is a four-sided spinning top with one Hebrew letter written on each side. Apart from being a part of the traditional children's game usually played during the Jewish festival of Hanukkah, dreidels have a deep spiritual meaning attached to it. Dreidels come in various sizes and materials, and they get their spiritual significance from the message written on them.

Dreidel shape has been a part of designs owing to its simplicity and significance. The perspective of shapes as a design element depends on one's viewpoint, and they have the power to control one's feelings. Dreidel shape as a design element is not constrained by color and offers a lot of variability in terms of shape. Designs with dreidels are modern and yet timeless at the same time.

Dreidel shape can serve as valuable design element in the hands of designers to create contemporary art with meaningful insights. In this chapter, designs incorporating dreidel shapes in different forms and colors are presented. The simple designs complemented by the vibrant colors will serve as a useful template for designers (Figs. 9.1, 9.2, 9.3, 9.4, 9.5, 9.6, 9.7, 9.8, 9.9, 9.10, 9.11, 9.12, 9.13, 9.14, 9.15, 9.16, 9.17, 9.18, 9.19 and 9.20).

K. Murugesh Babu et al., *Abstract Pattern Illustrations for Textile Printing*, Textile Science and Clothing Technology, https://doi.org/10.1007/978-981-16-5975-1_9

Fig. 9.1 Elongate: Illusion of creating slender width with a relative diamond Pattern

Fig. 9.2 Shaper: Cutting edge traversed across the design clubbing flat shape into a mesh abstract

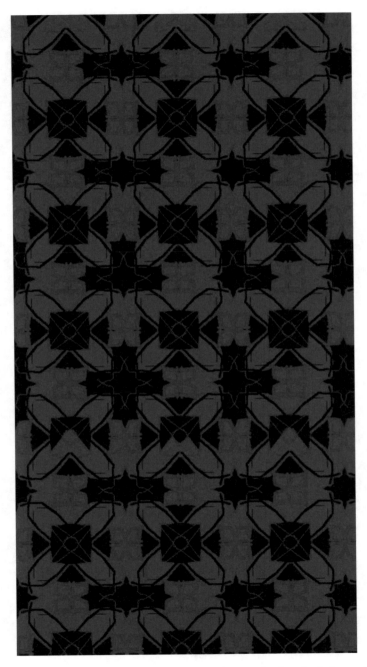

Fig. 9.3 Sundown: Candid evening shade below the horizon turning into the darkness of a night

Fig. 9.4 Duskiness: Dim non-luminous light with a contrary bright yellow creating a focused warm vibe

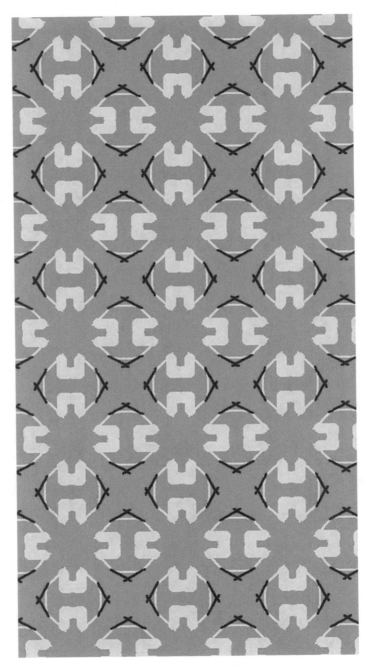

Fig. 9.5 Blitheness: A carefree pattern placed in a happy spirit with repeated Pointers

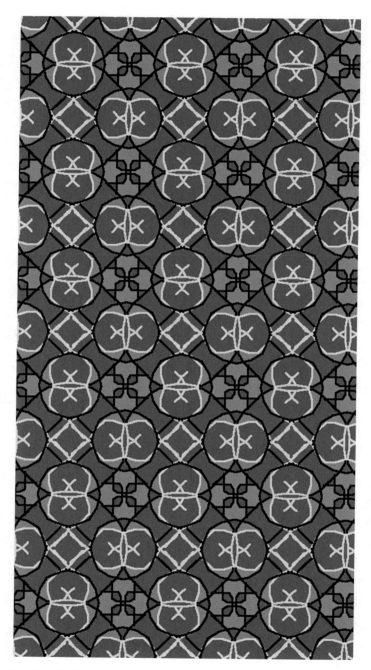

Fig. 9.6 Cessation: The horns and diamonds processed in a symmetrical way with white highlighter

Fig. 9.7 Objective: Colorful dreidels influenced with complimentary hues aiming at the center

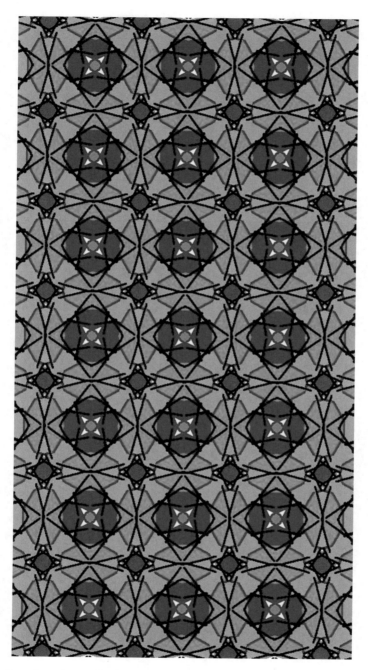

Fig. 9.8 Corundum: A crystalized arrangement formed with a variety of coarse shapes in contrast to color blend

Fig. 9.9 Gem rock: Clear and focused formation with a solid pattern-like expression of a jewel

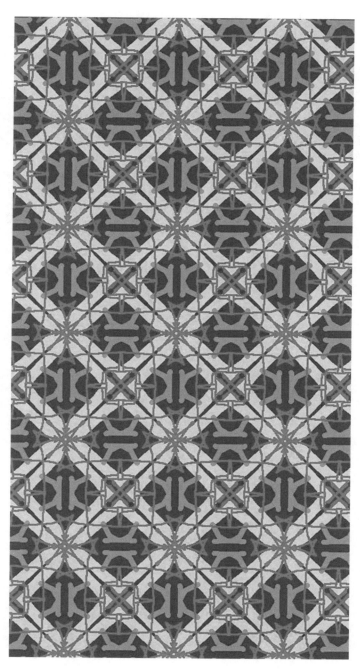

Fig. 9.10 Ice jewel: Coverage formation with linear shapes well suited for poised and elegant patterns

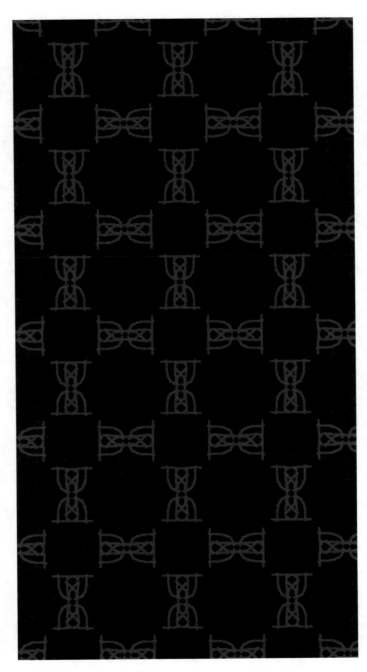

Fig. 9.11 Motif: Decorative design dominating a repeated color block pattern

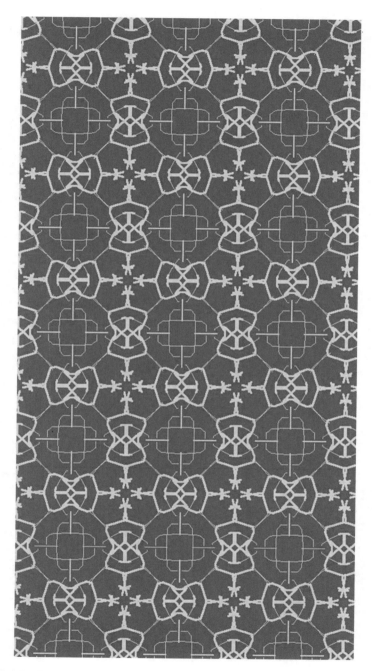

Fig. 9.12   Mindful: Focusing on the inclined shape creating a therapeutic and calm pattern

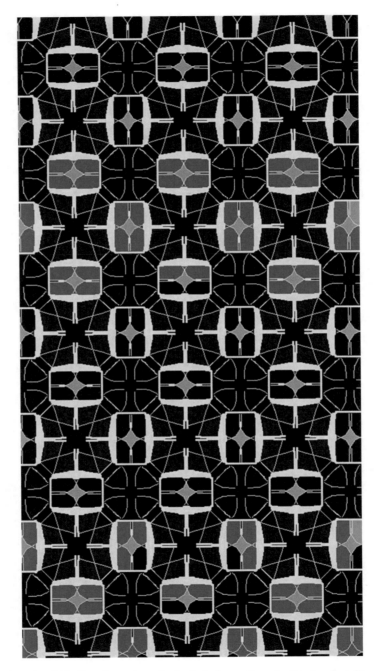

Fig. 9.13 Ellipsoidal: Transformation of spherical shapes through directional scaling into the design surface

Fig. 9.14 Monotonic: Modest and elegant monotonic artwork of interconnected dreidels

Fig. 9.15 Meticulous: Attention to motif detail creating a glazed and précised lead

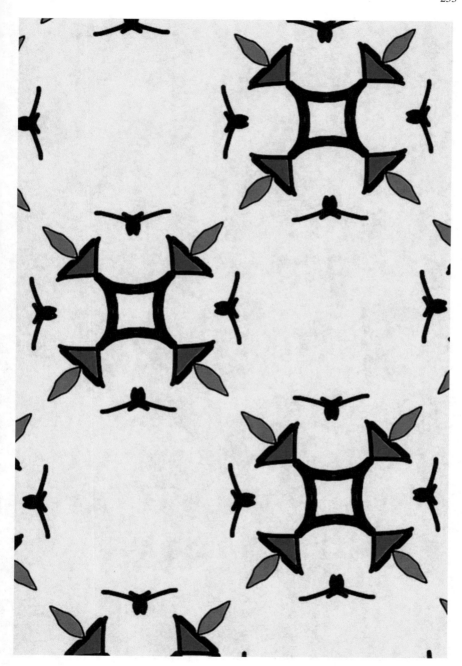

Fig. 9.16 Illuminate: A decorative quad symmetric motif illuminating the surrounding

Fig. 9.17 Objective: Back-to-back dreidels with complementary hues alternated by floral embedding

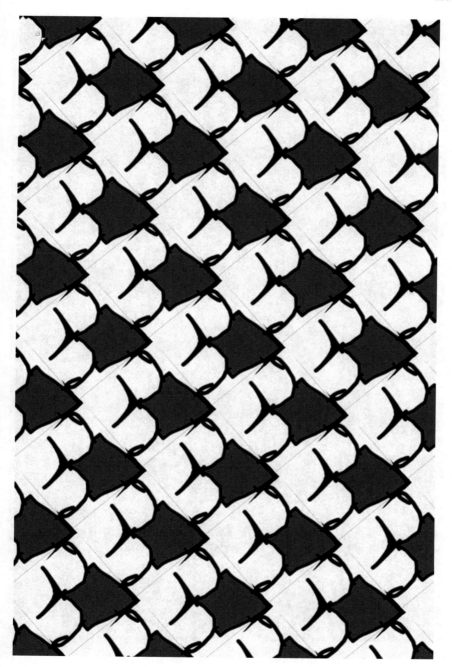

Fig. 9.18 Marine: A sophisticated and poise portrayal of marine life as a greyscale motif

Fig. 9.19 Sleek: Cyclical motif forming a binding interactive surface applique

Fig. 9.20   Furnace: Vivid tribal abstract with red and grey artwork

Chapter 10
Petals

Petals come in diverse sizes, shapes and colors. The brightly colored petals add beauty and splendor to the plant. They are the visually prominent elements in a plant placed to catch the attention of pollinators. It has been triumphant in not only capturing the attention of pollinators but also artists around the world. Petals come in an extensive range of colors and patterns giving ample inspiration to the designers to create exquisite designs.

The complex and ornamental arrangement of petals with their diverse pigmentation patterns has been an inspiration for artists for centuries. Designs inspired by petals are timeless and non-depreciating in value.

The use of petals as design elements in designs adds beauty, harmony and expression to the design and awakens the visual senses of the viewer. The vibrant colors and patterns of petals impart life to the designs and add energy that propagates through the design to the onlooker.

In this chapter, vibrant and vivid designs using petals have been presented. The colors and patterns used in the design will enlighten and facilitate the designers to create excellent designs (Figs. 10.1, 10.2, 10.3, 10.4, 10.5, 10.6, 10.7, 10.8, 10.9, 10.10, 10.11, 10.12, 10.13, 10.14, 10.15, 10.16, 10.17, 10.18, 10.19, 10.20, 10.21, 10.22, 10.23, 10.24, 10.25, 10.26, 10.27, 10.28, 10.29 and 10.30).

© The Author(s), under exclusive license to Springer Nature Singapore Pte Ltd. 2022
K. Murugesh Babu et al., *Abstract Pattern Illustrations for Textile Printing*, Textile Science and Clothing Technology, https://doi.org/10.1007/978-981-16-5975-1_10

Fig. 10.1 Hexahedron: Regressive hexagons shapes in cool colors giving a moving pattern

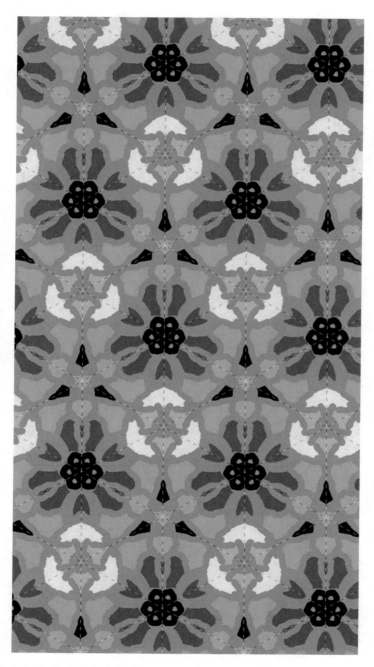

Fig. 10.2 Ethnologic: Tribal ethnic abstract artwork pattern with bright colors to infinity

Fig. 10.3 Vadalum: Primitive floral design in a dark background that is adorably and pleasantly simple

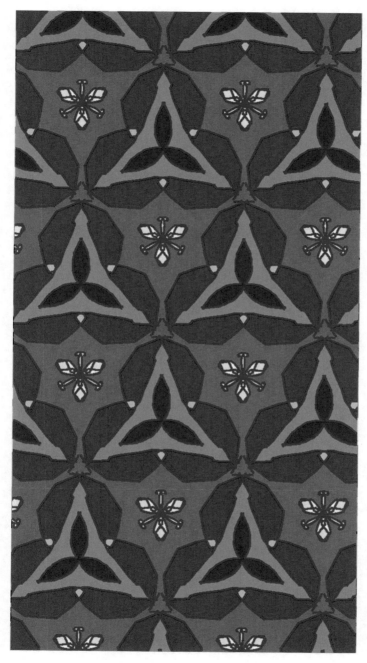

Fig. 10.4 Perception: Stunning contemporary abstract having subtle and warm colors

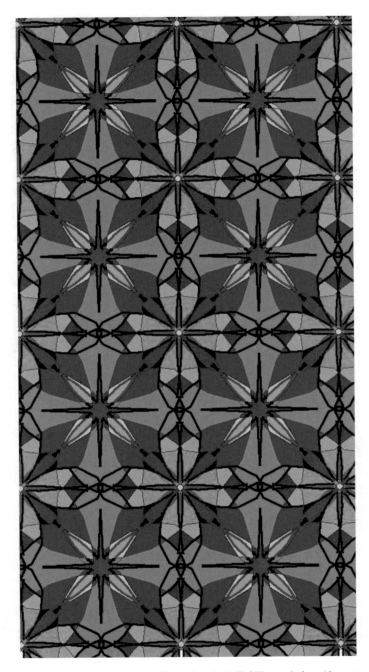

Fig. 10.5 Enlighten: Structured pattern enlightening the belief like a window Abstract

Fig. 10.6 Ratchet: Angled wheel in motion creating a bowtie shape in negative space

Fig. 10.7 Imprint: A multi-colored pattern encompassing the cool imprint into floristic abstract

Fig. 10.8 Spellbinder: The naïve amalgamation of abstract shapes creating a spellbinder pattern

Fig. 10.9 Simpatico: Pattern made with different abstract shapes in a closed concord making it look complete and friendly

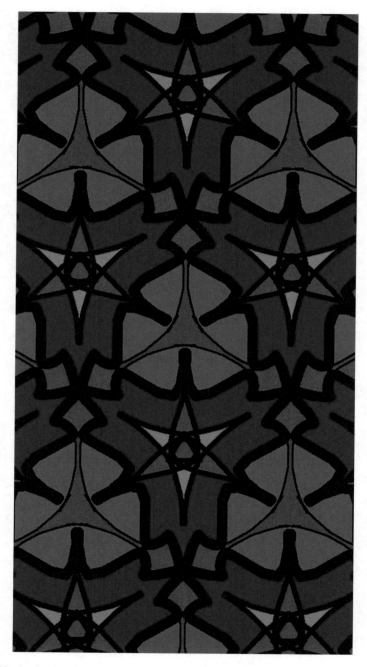

Fig. 10.10 Arctic: A polar print with striking color theory implies to a serene expression being cold and edgy

Fig. 10.11 Tessellation: An arrangement of abstract shapes closely fitted forming a state of surface art

Fig. 10.12 Mosaic: A bright polar extending the cool notion into the floristic pattern

Fig. 10.13 Frigid: The formal pattern created by linear style strikes a sheer elegant behavior

Fig. 10.14 Effervescent: Effervescent rows of floral crepes creating a pattern of bright angled slats

Fig. 10.15 Flourish: Yellow petals spread on a purple base and a brunette mesh

Fig. 10.16 Rainbow: Lovely rainbow-colored petals strewn with a purple periphery

Fig. 10.17 Classic: A classic structured pattern of floral mesh depicting the grace and style

Fig. 10.18   Magic: A mesh of floral hexagons portraying boldness and charisma

Fig. 10.19 Sapphire: Royal celestial hues sourcing to the vibrancy of the geometrical pattern

Fig. 10.20 Sea tide: Rising and falling of shapes combining to the bulging effect on table

Fig. 10.21 White swoosh: Flash of colors highlighting the wave like a sudden rush of air or liquid

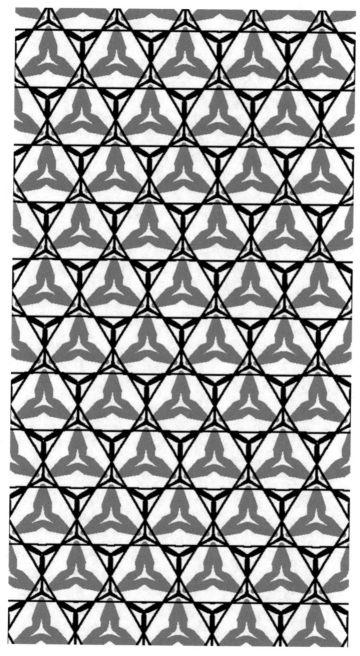

Fig. 10.22 Nettle: Pattern looks like herbaceous plant with jagged leaves covered with hexagon-shaped stinging hair

Fig. 10.23 Shrubbery: Equal sized perennial inverted and regular triangles showing negative and positive spaces in this abstract focused pattern

Fig. 10.24 Arachnid: Cluster of shapes forming insect imitating pattern denoting body and legs frame with white as face

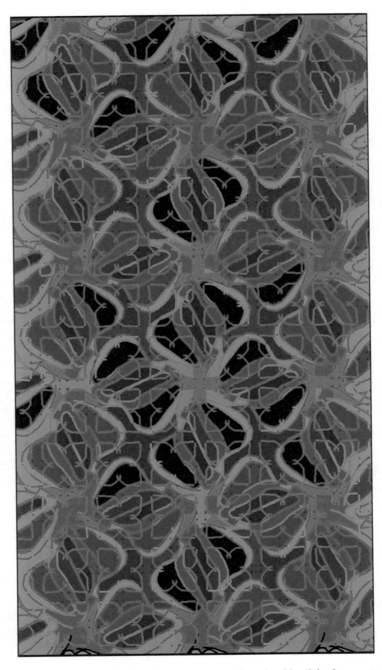

Fig. 10.25 Pastels: Abstract arrangement of Pastel colored petals with subtle clover patterns

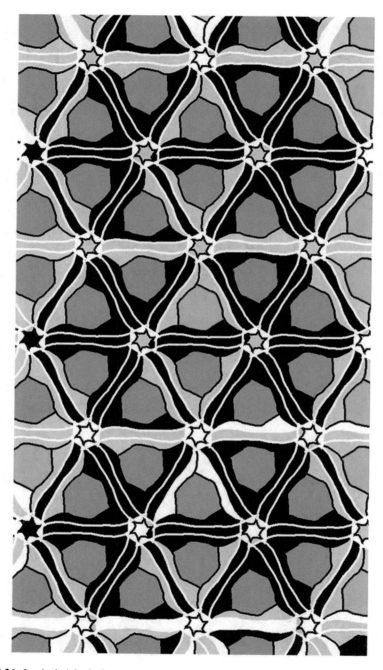

Fig. 10.26 Logical: A logical arrangement of bi-color petals in star-centered segmented hexagons

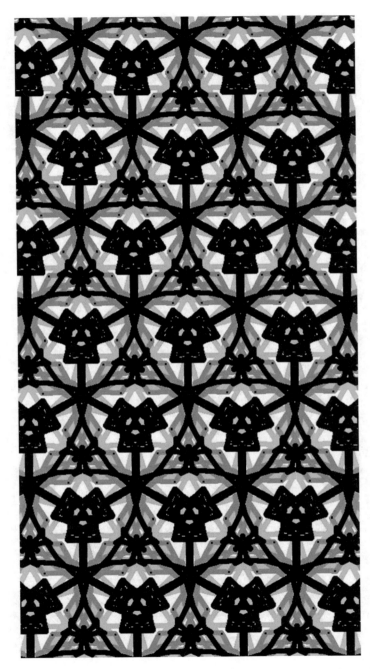

Fig. 10.27 Casement: Beautiful intricate meshwork looking like hollow molds for a window or any other vertical hinge in an architecture

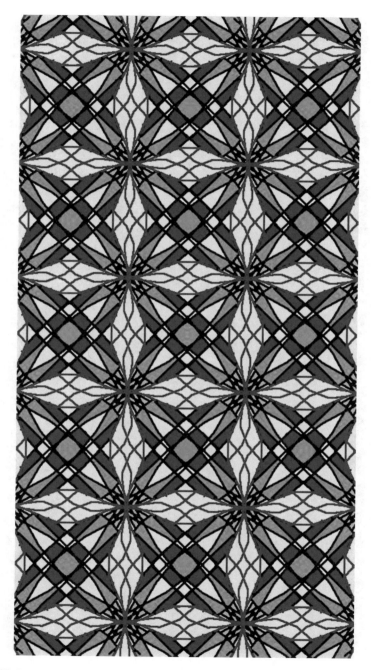

Fig. 10.28 Incarnated: Ancient dispersal of shapes well placed representing floral pattern with geometrical space

Fig. 10.29 Splash: Scattered geometric abstract shapes with a splash of colors

Fig. 10.30 Oval: A repetition of oval decorations with floral embedding in the center